Radical Adaptation

This book considers the everyday conduits through which climate instability is revealing itself: the storm sewer drain on your street, the powerlines transporting your electricity, the mix of vegetation in your backyard or neighborhood park – these are the pathways through which climate change is most likely to impact your life. For many, these are the last places we expect it to. In this book, Stone's aim is to understand how climate change is altering our lives in the present period – this period of transition between the ancient, stable climate of our ancestors and the unfolding, no longer stable climate of our children – and how our cities might adapt to these changes. Stone's concern is with the risks posed by a new environmental regime for which our modes of living are ill-adapted, and with how these modes of living must be altered – radically altered – to persist in a climate changed world.

Brian Stone, Jr. is a Professor in the School of City and Regional Planning at the Georgia Institute of Technology, where he teaches urban environmental planning and directs the Urban Climate Lab. Stone's program of research is focused on urban scale drivers of climate change and has been supported by the National Science Foundation, US Centers for Disease Control and Prevention, and US Environmental Protection Agency. Stone's work on urbanization and climate change is regularly featured in *The New York Times*, *The Washington Post*, and on *National Public Radio*. He is author of *The City and the Coming Climate: Climate Change in the Places We Live* (2012, Cambridge University Press), which received a *Choice* Outstanding Academic Title Award. Stone holds degrees in environmental management and urban planning from Duke University and the Georgia Institute of Technology.

Radical Adaptation
Transforming Cities for a Climate Changed World

BRIAN STONE, JR.
Georgia Institute of Technology

 CAMBRIDGE
UNIVERSITY PRESS

Shaftesbury Road, Cambridge CB2 8EA, United Kingdom

One Liberty Plaza, 20th Floor, New York, NY 10006, USA 477

Williamstown Road, Port Melbourne, VIC 3207, Australia

314–321, 3rd Floor, Plot 3, Splendor Forum, Jasola District Centre, New Delhi – 110025, India

103 Penang Road, #05–06/07, Visioncrest Commercial, Singapore 238467

Cambridge University Press is part of Cambridge University Press & Assessment, a department of the University of Cambridge.

We share the University's mission to contribute to society through the pursuit of education, learning and research at the highest international levels of excellence.

www.cambridge.org
Information on this title: www.cambridge.org/9781009211178

DOI: 10.1017/9781009211192

First published 2024

A catalogue record for this publication is available from the British Library

Library of Congress Cataloging-in-Publication Data
Names: Stone, Brian, Jr., author.
Title: Radical adaptation : transforming cities for a climate changed world / Brian Stone, Jr., Georgia Institute of Technology.
Description: New York : Cambridge University Press, 2024. | Includes bibliographical references and index.
Identifiers: LCCN 2023031394 | ISBN 9781009211178 (hardback) | ISBN 9781009211154 (paperback)
Subjects: LCSH: Municipal engineering. | City planning – Climatic factors. | Climate change adaptation. | Cities and towns. | Urban climatology.
Classification: LCC TD168.5 .S76 2024 | DDC 628–dc23/eng/20230822
LC record available at https://lccn.loc.gov/2023031394

ISBN 978-1-009-21117-8 Hardback
ISBN 978-1-009-21115-4 Paperback

For *Josie*

Contents

Acknowledgments

This book was born of an unexpected outcome. Interested to understand the pace of warming in cities relative to that of the planet as a whole, my research group – the Urban Climate Lab at the Georgia Institute of Technology – carried out an analysis of temperature change in the most populous US cities. The outcome of this study enabled a ranking of US cities based on the rate of urban warming, and the most rapidly warming city was not found to be a desert city or the largest city or the city with the highest rate of population growth. The most rapidly warming city in the United States at that time was Louisville, Kentucky – a mid-sized city in a humid-subtropical zone with ample access to water. The reasons for this accelerated warming trend are varied and range from a paucity of tree canopy in the downtown district to the upwind positioning of large, heat-emitting industries. To the immense credit of the city's leadership, rather than protest or ignore this rather ignominious ranking, Louisville committed to cool itself down.

In partnering with the Office of Sustainability in Louisville, in concert with a set of foundations helping to support the work, the Urban Climate Lab sought to explore the extent to which a large city could alter its own weather. Most critically, to what extent would a citywide increase in tree canopy and the use of cool roofing materials not only lower neighborhood temperatures but potentially reduce the annual number of heat-related deaths? Which residents are most vulnerable to these impacts? A decade-long journey to answer these questions – in Louisville and numerous other cities – has given rise to the ideas in this book.

I am most grateful to the array of one-time students, colleagues, planners, and forward-looking community advocates who have enabled the research and experimentation informing this book. This list includes most centrally Josh Bakin, John Bolduc, Christian Braneon, Christy Brown, Bumseok Chun, Matt Grubisich, Dana Habeeb, Dudley Hartel, Michelle King, Maria Koetter, Kevin

Lanza, Evan Mallen, Ted Russell, Jason Vargo, and Howard Wertheimer. This work has benefited greatly from partnering with colleagues on a series of academic papers that have informed this work, including the late (and much missed) Godfried Augenbroe, Ashley Broadbent, Matei Georgescu, Santiago Grijalva, Carina Gronlund, Sharon Harlan, David Hondula, Scott Krayenhoff, Larissa Larsen, Marie O'Neill, and Mayuri Rajput. Most recently, I am indebted to Atlanta city council members Liliana Bakhtiari and Matt Westmoreland for their support of climate vulnerability work in Atlanta, which informs the book's Postscript.

A sincere thanks to Matt Lloyd and his team at Cambridge University Press for enthusiastically taking on another unconventional book on climate change. I am further grateful to my colleagues in the School of City and Regional Planning at Georgia Tech who supported this book through providing the needed leave time to complete its writing. A special thanks to Michaela Master for assistance in compiling the book's set of references.

Last, I am ever grateful to Josie, Wesley, William, Vivian, and Owen for sharing me with this work. You are my hope.

Prologue: Dead Pool

In the prior age, it was sometimes referred to as the Saudi Arabia of water, a nation so amply endowed with freshwater – in the form of the Earth's most massive river, sequestered within the planet's largest expanse of virgin rainforest, and elevated aloft into the wind currents linking inland jungle to ocean – it rivaled the world's petrostates in terms of sheer liquid wealth. Accounting for less than 2 percent of the planet's land area but by some estimates 20 percent of its total freshwater, with none of it stored in shrinking glaciers, Brazil was an unlikely setting for the armed hijacking of water trucks. But many months into a deepening drought in 2014, even military escorts were failing to ensure the safe passage of water deliveries to parts of São Paulo, a megacity of 21 million edging ever nearer to the largest forced evacuation in history [1]. As the Paulistanos were learning, even an abundance of natural resources was no match for an increasingly unnatural climate.

The first residents to riot were in the suburban city of Itu. While city dwellers in central São Paulo were working to manage periods of a day or two with no flowing water from taps, the supply in Itu was far less frequent, leaving some residents without access to water for weeks at a time [2]. Many months into the nation's most severe drought on record – with less rain falling in the 2014 and 2015 rainy seasons than in a typical dry season – regional reservoirs were highly depleted, in some cases falling to 2 percent of capacity [2]. Even if sufficient pressures remained to pump from the final shallows of a depleted reservoir, known in technical parlance as the "dead pool," the silty remnants were often too contaminated for drinking. With no regional water available, the only option was to truck in water to distribution sites – a strategy far unequal to demand and vulnerable to armed conflict.

Of all the societal benefits provided by collective settlement in cities, clean drinking water is perhaps the least appreciated of the modern world. Dating back to the ancient Minoans of Crete, who were the first to transport water over significant distances and distribute to urban settlements via aqueducts, urban water delivery systems – now coupled with wastewater removal – remain the

1

most fundamental technology of cities [3]. While local storage of solid waste enables trash removal in cities to operate on a weekly or biweekly schedule, and millions of city dwellers even today live without access to electricity, nowhere in the world can a large population be sustained for more than a few days without access to potable water. In periods without it, all other human concerns become secondary.

Paulistanos are not able to draw drinking water from the sizable rivers that flow directly through their city. With few prohibitions on what can be discharged into the Tiete and Pinheiros rivers, ranked among the most polluted waterways in the world, every conceivable byproduct of an industrial megacity is entrained in their flow. In total, an estimated 4 million residents of São Paulo lack access to basic sanitary sewage systems, with most of this human waste finding its way to the rivers. So complete is the degradation of the Tiete and Pinheiros that subsurface blockages to stormwater pipes can only be removed by divers in fully enclosed, deep-sea diving suits. The bounty of such dives reads as though drawn from the pages of a shopworn novel: discarded guns and knives, a suitcase filled with cash, a dismembered body [4].

So, the city must look elsewhere for a viable source of drinking water. Developed in the 1960s, the intended solution was a chain of large reservoirs to the north known as the Cantareira System. Designed to meet the population growth projections of that era, and calibrated to rainfall patterns of the earlier, stable climate, the reservoir system, while among the world's most extensive, has increasingly failed to provide an adequate reserve of water for São Paulo (Figure P.1). By 2014, lower than expected rainfall from a single rainy season was sufficient to bring the water supply for a city of 21 million inhabitants within 100 days of depletion [5].

Rainfall, of course, is only one of several variables that govern the life of a city's water supply. The rate at which water is withdrawn from a reservoir is equally determinative of how many days of drinking water remain and, on this front at least, São Paulo seemed to hold a salvageable hand. The most promising card to play was a reduction in the rate of regional water withdrawals for agricultural and industrial uses, estimated at the time to account for more than 90 percent of the total water usage in the state of São Paulo [6]. This statistic alone is worth considering: Less than a tenth of the drinking water system constructed to support the world's fourth largest city was used to address the water needs of the city's residents, with the remainder flowing to agricultural uses (much of the harvest destined for export), manufacturing, and power plants. The first water rationing to occur, however, would not be in an industrial zone but in outlying residential neighborhoods of the metropolitan region, and it would come without warning.

Figure P.1 Pumping water from the dead pool in the Cantareira reservoir system, February 2015. Paulo Whitaker/Reuters.

A second lever to pull when confronted with a mounting water shortage is the reduction in water losses from the vast network of tunnels and pipes delivering water from the reservoir to the tap. While all major urban water systems suffer from unintended leakage, São Paulo excels in this area, with almost 40 percent of the water that is stored and treated lost to failures in the subterranean distribution system [7]. Here again, a moment of reflection seems warranted: More than a third of the vast quantities of water collected and stored in the South America's largest water system – a system that requires prodigious amounts of energy to treat and deliver drinking water – never makes it to a tap or a hose or an irrigation system. São Paulo loses daily a quantity of water that would meet the total demand of all but the largest cities in North America. An emergency campaign to identify and repair the largest of these system failures could have measurably extended the life of the dwindling reservoir system but was not undertaken during the crisis.

With two viable options to extend the life of the region's reservoirs, state and local officials decided on a third course of action: an unannounced rationing of water that would most severely impact the poorest residents of the city. Referred to as "hidden rationing," the water utility substantially lowered water pressure in the system, initially at night but later for days at a time, which zeroed out water access entirely for residents at the furthest reaches of the distribution system where, typically, the poorest families reside [8]. As neither the water utility nor local or regional government officials acknowledged this decision, and, in fact, regularly denied it, there is no basis to assess their reasoning. But one variable seems likely to have played a role: The deepest drought in Brazil's history had arrived on the eve of the 2014 World Cup, with the opening match to be played in São Paulo only a month after the first taps ran dry.

The first to take to the streets were women. Disproportionately responsible for household labor, women were bearing the brunt of the water crisis. From the vantage of continuously flowing household water, it is easy to overlook the tremendous effort required to locate, gather, transport, and store water for daily use. For starters, information on where water can be obtained – from a newly drilled well or water tanker truck – was infrequent and unreliable, given the government's reluctance to broadcast the shortage. So, trips to gather water typically included a long wait in line with little guarantee of sufficient supplies to fill a vessel. If a source is secured, water ranks among the heaviest liquids found in nature (water is heavier than crude oil, for example), greatly limiting the quantity that can be transported over any measurable distance. Last, few households served by a municipal drinking water system have water storage capacity available, resulting in the need for continuous trips to collection sites.

But additional labor is only the beginning of how a household must adapt to limited water – equally important is how the water is used. Women tend to be the first to ration their own consumption, as they prioritize scarce drinking water for children and other family members. The result for many women in poorer areas of São Paulo, where flowing water was ceasing for days at a time, was a chronic state of dehydration. Soon to follow was a sharp increase in urinary tract infections, as the passage of water through digestive systems is essential to the flushing of bacteria. For others, more severe gastrointestinal illnesses were tied to the consumption of contaminated water from hastily drilled wells [9].

It has been observed that in periods of catastrophic weather, such as in the aftermath of a hurricane or wildfire, communities come together. But in periods of resource depletion, such as a famine or drought, communities come apart. Many weeks into severe water rationing in São Paulo, water-starved communities were coming apart. Consistently assured by local politicians that there was no water shortage, that steps were being taken to ensure a steady delivery of drinking water throughout the drought – even as no water flowed from household taps for days running – residents who had for months exhibited resilience began to view the crisis as much as an instance of governmental failure than as an extreme weather event. As with all riots, the Brazilian water riots were born not just of a crisis but of a failing governmental response to crisis.

Footage from the initial protests show familiar actions of civil resistance – marchers with signs and banners, women sitting in streets to disrupt traffic, a police presence on the periphery. With little governmental acknowledgment of what were becoming daily protests, more aggressive tactics ensue. Streets are barricaded and then the barricades set afire. Rocks are hurled at public buildings. And the police, now in full riot gear, fire tear gas to disperse the crowds. Soon to follow is the looting of stores in search of bottled water and the ultimate armed hijacking of water trucks [2]. Less than six months from the official declaration of a drought, the largest city in Brazil and a global model of rapid modernization for the developing world had lost the capacity to deliver water to its population. Beyond the reach of modern technology, globally available capital, or human ingenuity, the system had failed to provide its most basic and ancient ingredient: the regular return of rain.

Several years after the drought, in 2018, the newly elected president of Brazil would proclaim an intent to accelerate development of the Amazon. Elevated to power on a political platform emphasizing further global integration of the Brazilian economy and the promise of continued rising standards of living, President Jair Bolsonaro sought to relax a suite of rainforest protections put in place over the prior decade [10]. Alarmed by the increasingly rapid rate of deforestation in the Amazon over the 1990s, a period in which several hectares of forest were being cleared every hour, earlier governments had put in place reforms that reduced the rate of deforestation by an estimated 70 percent between 2005 and 2014 – a trend that would sharply reverse under Bolsonaro [11].

With limited roads constructed into the interior of the remaining Amazon – a still vast area the size of Australia and largely populated by Indigenous groups – illegal loggers transport long-blade chainsaws via motorcycles along circuitous networks of footpaths. The direct product of the clear-cutting can carry significant value, but little if any of the felled wood from the Amazon's interior will make it out of the jungle. With no viable route for transport, the cleared forest tracts are simply burned on-site, extinguishing in the process an unknowable extent of animal life and releasing vast quantities of stored carbon to further warm the atmosphere. The intended export of the illegal but increasingly sanctioned logging in Brazil is not wood but beef, with the planet's most essential and teeming ecosystem rapidly degraded to a near-silent monoculture of grazing cattle [10].

The process of converting rainforest to beef is among the most celebrated in the annals of ecological mismanagement. For at each step in this process, energy is wasted. The amount of solar energy that can be converted by the rainforest into edible plants, such as jackfruit, to take one example, is much greater than the energy that can be converted into plant protein in the form of grass. As a result, the rainforest-for-pastureland tradeoff yields less usable energy for humans, as there is simply far less biomass in an acre of grass than in an acre of dense and multilayered rainforest.

Were it possible for humans to consume the pastureland's grass directly, we would need about three acres of pasture to provide an equivalent amount of caloric energy as found in a single acre of rainforest. Put another way, we lose about two-thirds of the available caloric energy when we convert rainforest to pasture [12]. Yet it is the next step in the process – the conversion of grass to beef – that really distinguishes Amazonian cattle ranching. Cows may be the least efficient machine yet conceived by nature and industry for the conversion of biomass to animal protein, as so much of the energy content of the grass is needed simply to propel the animal's daily wanderings. If 66 of the 100 units of potential solar-derived caloric energy are lost in converting rainforest to pastureland, roughly another 32 units are lost in converting pastureland to beef [13]. So, to sum up the math, about 2 percent of the caloric energy captured by an acre of rainforest can be converted into cow muscle through deforestation – 98 percent of the captured solar energy is lost. These numbers are borne out by the average density of cattle ranching in the Amazon today, with about 2.5 acres of grazing land required to sustain a single cow – enough land, if planted in crops, to feed a family year-round [14].

Vegans delight in highlighting the immense ecological costs of meat production, with one analysis finding the consumption of a single Brazilian hamburger to carry the carbon footprint of a 200-mile road trip [15]. But an additional dimension of Amazonian mismanagement is often omitted from meat-shaming computations: The rainforest is not simply the beneficiary of plentiful rain; it is itself a reservoir from which the rainwater is drawn. And the reservoir is massive, with parts of the Amazon receiving more than 400 inches of rain each year (the United States, by comparison, averages 30 inches of rain per year) [16]. The combination of saturated soils and intense solar gain due to the forest's equatorial location translates into some of the highest levels of evapotranspiration (the evaporation of surface water plus the release of water vapor from green plants) measured on Earth. This invisible, rising river of water vapor generates clouds across the Amazon and throughout much of southeastern Brazil, a largely self-contained hydrologic cycle through which the forest seeds its own rain and, ultimately, the rainfall received by urban areas downwind.

Cattle grazing requires extensive grasslands to support it, and it equally requires rain. The challenge moving forward for Amazonian cattle ranchers is that the provision of the first input directly diminishes the second. With each acre of rainforest felled to clear more land for grazing, the reservoir of moisture that drives the regional hydrological cycle is, much like an aquifer, drawn down. The impact of this drawdown is a regional climate that is becoming warmer and drier over time, with less rain falling today than several decades ago. Climate models consistently show that continued destruction of the Amazonian rainforest will further diminish rainfall over time, substantially under some scenarios [17]. And so, the direct connection between the area of rainforest and patterns of regional rainfall presents a clear dilemma to Brazilian leaders. To sustain itself, Amazonian pastureland requires what remains of the rainforest for natural irrigation. The same is true for the city.

While the drivers of the deepest Brazilian drought on record are difficult to quantify with any precision – in addition to a gradual drying of the Amazon an active El Niño cycle is thought to have played a role – the implications for the entirety of the São Paulo population were growing ever more acute by early 2015. With less than half of the typical rainfall occurring in the 2014 rainy season, the subsequent rainy season also fell short of historical averages, yielding little recharge of the region's depleted reservoirs. A few months into 2015, the rationing of water that had principally impacted outer-lying areas of São Paulo was expanded to the most heavily populated zones, resulting in no flowing water for days at a time. The water utility proposed in February a rationing system unlike any previously seen in a major global city: Without the return of regular rainfall, no water would flow to the city's homes for five days each week [18].

As the frequency of extreme weather events is creating ever more prolonged breakdowns in the critical infrastructure of large cities, familiar patterns of urban life are altered in respects both profound and mundane. Still lacking electrical power several weeks after Hurricane Maria, in 2017, San Juan, Puerto Rico had all but vanished from nighttime satellite imagery, with no power to operate streetlights and the other innumerable sources of illumination across a densely settled area (Figure P.2). But what was most arresting about the stricken city, according to those who visited at the time, was not the absence of light but the eerie, deeply unsettling silence of a major metropolis rendered inoperable. In a similar sense, what most clearly foretells the advance of a wildfire into an urban area is not the blaring of sirens or the beating of helicopter blades but a soundless, light-altering haze in the sky, increasingly intense and oddly beautiful.

Figure P.2 Blackout in San Juan, Puerto Rico in the aftermath of Hurricane Maria,
September 20, 2017. Alex Wroblewski/Getty Images.

For the city without water, what most characterizes the urban landscape is
neither a silencing of the streets nor an altering of the skies. What most
characterizes a city without water is the ubiquitous, unrelenting odor of
human waste. Many months into the deepest water crisis to confront a city of
more than 10 million residents, Paulistanos were left with little choice but to
triage the dwindling quantities of water available. The lion's share would be
reserved for drinking and cooking, with some remaining for the washing of
hands and faces – or perhaps the soaking of long unwashed clothes. Lower on
the list, due as much to the large quantity required as to the perceived imperative, was the flushing of toilets.

For many, it was this stench more than a scarcity of drinking water that was
the most startling aspect of the crisis. While nourishment could be found for
most in the form of bottled or stored water, an inability to bathe oneself or flush
away waste was perhaps most revealing of the mounting vulnerability of urban
life in a climate changed world. Even the most basic of urban services – the
provision of shelter, water, and waste removal – could not be relied upon in an
unstable climate. As observed by one public official, "Water is about human
dignity . . . When people can no longer wash themselves or use the toilet or take
care of their children, they start to panic" [19].

Compounding this panic was the regular onset of blackouts. Endowed with such an abundance of freshwater, Brazil has extensively dammed many of its large rivers for the production of hydropower, accounting for more than 80 percent of the country's electricity generation [20]. As reservoir levels dropped, the capacity for hydropower stations to produce electricity was also diminished, creating disruptions in the electrical grid. With a deepening drought, even coal or gas-fired power plants are impacted, due to their reliance on river or lake water for the cooling of boilers, which further undermines the resilience of an electrical grid during periods of water scarcity. Water, in this sense, is not simply one of several elements that sustains urban populations; it is the central element. Without it, the city quickly reverts to a preindustrial condition – a condition that was never contemplated for millions of inhabitants.

For most reading these pages, the experience of a blackout or a water main break, leaving one without power or running water, is familiar. Fewer likely have experienced in a major city both conditions simultaneously and, if so, for more than a day or two. Perhaps more than any other threat we can imagine from a world with erratic weather, including the violence of an intense hurricane or wildfire, the failure of critical infrastructure systems in urban areas is not only far more probable in the coming years but it is arguably more dangerous, for it carries with it the breakdown of civic order.

The cone of a forecasted hurricane track is known in advance; its path of devastation established and unchanging once the storm has passed. In the aftermath of an acute weather event, governments are enabled to take action, implement a plan to rebuild, and project a sense of civic control. The same is not true for a drought-induced infrastructure failure, which lacks easily implementable solutions or a certain timetable. At what point should a megacity of 21 million inhabitants be evacuated and, if even possible, where would the population be sent? In the early months of 2015, São Paulo was experiencing not only an ecological collapse but a collapse in the veneer of social control that is as fundamental to the operation of cities as drinking water and waste removal. For Paulistanos, the climate crisis had arrived not in the form of a violent storm surge but as a revelation of institutional incapacity.

Unable to resolve a crisis rooted in both long-term environmental and infrastructure mismanagement, at the local and global scales, governmental officials prepared for the only crisis for which action was possible: the rioting of its own population. As described candidly by one official:

> We were desperate. The reservoir level was just going down and down. We knew that when people don't have water, they go crazy. We had seen the protests in smaller

cities where people were breaking into property to steal water. We imagined what they would be like here with 21 million people. We thought about the hospitals unable to treat patients and children having to stay home from school. It would be chaos [21].

The first priority was ensuring the continuing function of critical institutions – buildings no less dependent on water for operation than residential zones. To ensure continued water delivery, utility workers constructed a new, parallel distribution system directly connecting to emergency water reserves the 500 most important buildings across São Paulo. Large hospitals, critical governmental buildings, police stations, and prisons were all to receive continuing water supplies, even as the rest of the city ran dry. The remainder of the city's residents would need to rely on the limited water that could be supplied via tanker truck.

Another of these buildings was a newly constructed emergency operations center. Such a center was already in place, but its location was publicly known. Initially intended to direct preparations and recovery operations for natural disasters, the focus of the emergency operations center had changed entirely, with the population to be safeguarded from danger having now become the danger. Members of the Brazilian intelligence services were quietly flown to the United States to train alongside Special Weapons and Tactics (SWAT) paramilitary units and returned with a newly contracted fleet of military assault vehicles [19]. Plans were made for a forced evacuation [22].

What was starting to unfold in São Paulo in the early months of 2015, as the taps ran dry for millions upon millions of residents who had not been adequately apprised of the depth of the crisis, or of the impotence of their government in the face of it, was something as of yet unseen in the modern era of human settlement: the staged depopulation of a city that remains physically intact. While the water utility would manage to deliver an absolute minimum quantity of drinking water to a majority of residents until the return of rain, resuming later in the year, the steps undertaken by the government in preparing for widespread rioting and a forced evacuation – the only administrative levers left to pull – attest to the inarguable reality confronted today by cities across the planet: Large-scale urbanization is only possible in a stable climate. Without it, the principal promise of a city – to make its residents less vulnerable to hunger, disease, crime, and ignorance – is unkept, serving to diminish rather than to enhance human welfare.

And it is the belief in this promise, still retained by much of the modern world, that has sustained cities until the present moment.

This is a book about transformation. With the workings of the global climate more responsive to human activity than at any time in the past, not only is an unrelenting trend toward hotter and unstable weather more predictable but so too is the direction of human history. Perhaps at no time in the past have the crises to confront humanity years into the future been more apparent, more clearly mappable, than at the present moment. If this statement is true, then it is reasonable to expect that the seeds of these crises have begun to take root among us and are for the first time visible to the naked eye. It is the transition from one environmental regime to another – a shoulder season between yesterday's climate and tomorrow's – that most clearly describes our present reality and announces the changes to come.

Although well underway, the transition between the old environmental regime and new operates on two very different timescales. At the scale of planetary change, our human-enhanced greenhouse effect is altering the biophysical composition of the Earth at a pace unprecedented in the paleoclimatic record. When plotted over geologic time, the rate at which carbon dioxide is accumulating in the atmosphere, paired with carbonic acid in the global oceans, assumes the shape of a vertical line – the only such verticality in more than 10 million years of reconstructed trends. Christened the Sixth Mass Extinction, the rate of species loss unfolding around us is on par with prehistoric meteor strikes and periods of violent, sun-blocking volcanism, with three-quarters of the planet's animal species – representing untold millions of years of evolution – presently disappearing over the course of a single human lifetime.

Viewed from the window of those living today, however, these changes can appear gradual. Yes, the spring flowers are blooming earlier than in one's youth, but a seasonal shift of a few weeks is hardly significant over the period of a fifty-two-week year, even if we are the first human generation to observe it. Tethered to an experiential timeline typically reaching only from our grandparents to our grandchildren, it is the period over which our bodies age, our children grow, and our cultural artifacts persist that most directly frames our temporal vantage – a set of waypoints too few to map the pace of planetary change. Of all the threats posed by a changing climate this is perhaps the most pernicious: The unprecedented rate at which the planet is changing has remained just outside of our biological capacity to detect it. But this is changing.

Throughout the entirety of the planet's history, a discontinuity between the pace of environmental change and the capacity of Earth's organisms to perceive it was of little consequence, as no single species was in a position to alter the composition of the atmosphere or the volume of the oceans. With this ecological law no longer in effect, the capacity for humans to recognize their impact on the global environment has assumed existential significance for all species. Ill-equipped by evolution to respond to slow moving threats, we require a nonexperiential mode of detection to better align the pace of planetary change with our perception of it. The mode of detection we have fashioned for this purpose is *science*.

But our science is failing us in the arena of climate change. Well constructed to look beyond our narrow temporal vantage – both backward and forward in time – the scientific method is less well suited to the purpose of rapid societal change. The rules of empiricism foundational to science – its need for extensive data collection and reproducibility – require time to satisfy. The larger the phenomenon, such as the workings of the global climate, the more time is needed for testing and validation, and the further along the climate change trajectory we advance.

Equally problematic is that the scientific method is silent on the question of how its advancements are to be communicated beyond a small community of specialists. Rigid in its formula for the production of new knowledge, science provides few insights for the essential task of knowledge translation. This task is left up to news outlets, advocacy groups, politicians, and, increasingly, our social media accounts. This is a problem.

With no uniformly accepted mechanism for societal translation, the journey from scientific certainty to popular understanding is often unscripted – particularly when confronted with well managed and funded misinformation campaigns. While prior public health battles, such as that over tobacco use, have yielded insights into effective risk communication, such episodes have been no less instructive in the dark arts of public deception. We have yet as a society to conclude that willful, premeditated misinformation campaigns that contribute to loss of life on par with military conflicts or pandemics are deserving of criminal sanctions, or even social opprobrium.

While contemporary debates over climate change have rendered the phenomenon complex and, to some degree, exotic in the eyes of the general public, it is neither. Relentlessly characterized as unfolding in arenas distant from our everyday experience – the disappearing ice caps, the melting tundra, an unprecedented heat wave on another continent – the impacts of climate change most threatening to the majority of the global population are presently playing out in our own cities and neighborhoods. The storm sewer drain on your street, the powerlines transporting your electricity, the mix of vegetation and wildlife in your backyard – these are the conduits through which climate change is most likely to impact your life. For many, these are last places we expect it to.

In this book, my aim is to understand how climate change is altering our lives in the present period – this period of transition between the ancient, stable climate of our ancestors and the unfolding, no longer stable climate of our children – and how our cities might adapt to these changes. My concern is with the risks posed by a new environmental regime for which our modes of living are ill-adapted and with how these modes of living must be altered – *radically altered* – to persist in a climate changed world.

My use of the word "radical" in this book has a more specific meaning than in popular usage. Rooted in the general precepts of *radical planning theory* [23,24,25,26,27] – an orientation to collective decision-making and action that operates outside of established governing institutions – the idea of radical adaptation recognizes that established modes of environmental management are failing to sufficiently protect urban populations from the extremity of climate impacts already unfolding. In the following chapters, I propose four elements of radical adaptation that represent a departure from conventional modes of urban planning, environmental management, or natural hazard mitigation. In place of our traditionally centralized approaches to environmental management, infrastructure for radical adaptation must be spatially dispersed. Rather than directing public resources to the areas of greatest potential physical impact, a radical approach to climate adaptation is directed to the zones of greatest human vulnerability. Novel strategies once perceived as socially untenable will need to be embraced. And lastly, under a radical framework, planned retreat within cities is not the last step in a series of adaptative actions but the first.

While the principal focus of adaptation will vary with geography, all cities must prepare by degree for a few universal elements of rapid ecological change: extreme heat, rising water, and prolonged droughts. Compounding these imminent threats to urban populations is the capacity of critical infrastructures to anticipate and rebound from systemic failure. Based on an assessment of these basic dimensions of adaptation – temperature, water (too much or too little), and critical infrastructure – a process of physical deconstruction within the most vulnerable zones of cities will be required. In many cities, this process is well underway.

Temperature, water, human technology. These basic elements of urban settlement are no less fundamental to the workings of our current society than in the earliest stages of the human project. And our cities are today no more resilient.

1

Heat

That the railway transport cars were labeled "SS" trains was only the first of several unfortunate parallels with an earlier forced migration of comparable scale. In what would be the largest domestic evacuation of a displaced population since World War II – and perhaps in human history – India's Shramik Special trains would transport an estimated 5,000,000 migrant workers over the period of a few weeks in May 2020 (Figure 1.1) [1]. The need to return a large population of migrant laborers from India's largest cities to their home states during a pandemic lockdown would necessitate extreme measures. The need to do so during one of the most intense heat waves on record would render the extremity of conditions on these trains unimaginable to anyone reading these pages. With temperatures in some regions of India exceeding 122°F [2], and air-conditioned railcars a luxury unavailable to most laborers, the train interiors were approaching, quite literally, the temperature of an oven. The total number of deaths from heat exposure is not known but press accounts reveal a regular removal of bodies from the SS trains [1].

While India is widely recognized as exhibiting a unique vulnerability to heat waves, its cities do not rank among the hottest on the planet. Desert cities of the Middle East, Africa, North America, and Australia all experience higher summer temperatures than the hottest of Indian cities. What renders Indians uniquely vulnerable to heat is not the extremity of temperature alone but a combination of very high temperatures and very high humidity. Like a winter coat in summer, humidity does not elevate the air temperature; it impedes the efficiency with which our bodies shed heat – amplifying the effects of temperature. A comparison of record temperature and heat index values, measuring the combined effects of heat and humidity, reveals the potential extremity of humid environments. The highest temperature yet recorded on Earth – 134°F in Death Valley, California – is about 10 degrees *lower* than

Figure 1.1 Laborers on Shramik Special train in West Bengal, India, May 24, 2020. Sanjoy Burdwan/iStock.

maximum heat index values measured during a recent heat wave in Delhi [3,4]. For Indians, humidity is the sharp end of the heat spear.

Highly adaptable to a wide range of climatic conditions across the planet, humans are surprisingly vulnerable to the most minimal of shifts in our internal physiologic state, with a rise or reduction in our core body temperatures of just a few degrees sufficient to induce death from heat or cold. Our principal adaptations are non-physiologic – buildings insulated against extreme temperature fluxes; clothing adjusted to the season; a cold shower on a hot day. When these external adaptations fail to guard against a swing in our core body temperatures of just a degree or two, our bodies deploy a small number of physiologic responses to maintain thermal equilibrium. In response to a falling core temperature, we shiver. Produced through a rapid tensioning and relaxing of skeletal muscles, shivering leverages the body's energy stores not for the usual purpose of carrying out work but rather for the waste product of muscular action: heat generation.

In response to a rising core temperature, we sweat. And, here again, in sweating the body makes strategic use of what is typically a waste product to maintain thermal equilibrium. Acquired directly through the ingestion of liquids and indirectly as a by-product of cellular respiration (the chemical processing of food energy), excess water is routed to the bladder for excretion. When conditions warrant, however, excess water can be excreted to the skin via sweat glands to protect against the incoming heat from the Sun and atmosphere.

While the water itself does not provide a cooling effect, its evaporation into the surrounding air diverts heat energy from the Sun and atmosphere from warming the skin to bringing about a phase change in water. Just as a boiling pot of water will maintain a constant temperature of 212°F as long as water remains to fuel a phase change to water vapor, moistened skin can suspend a rise in temperature with the continuous excretion of sweat.

Among the most powerful tools devised by the biophysical world to maintain homeostasis in mammals, or tolerable temperatures within ecosystems, this phase change in water exploits the Earth's natural conveyance system for cycling water between the biosphere and atmosphere to also cycle heat. Only when the water vapor cools and condenses back into liquid water, which occurs by rising to a higher altitude in the atmosphere, will the heat energy entrained within the vapor be released again to the environment – far removed from the vulnerable surface of the skin on a hot day.

High levels of atmospheric humidity are an acute threat to humans in hot weather not because humidity elevates the temperature of the air but because it impedes the efficiency with which water can evaporate from the skin's surface. Like a bowl of water filled to its rim, any further addition of liquid will cause the pool to overtop the vessel. As the atmosphere approaches 100 percent humidity, there is simply no additional capacity to absorb more water vapor. Under these conditions, water secreted by sweat glands can only pool on the surface of the skin, running off without extracting much heat from the surrounding air and failing to arrest a rise in core body temperatures. Once core body temperatures surpass 104°F, heatstroke – a rapid-acting and life-threatening condition – becomes likely [5].

Heat stroke is so named because it results from a starving of the brain of blood. Much like the more common ischemic stroke, resulting from a blood clot in the brain, a heatstroke also depletes the flow of blood to the brain – not from a physical blockage but from a massive diversion of blood to the body's extremities in a hypothalamic effort to dissipate heat back to the atmosphere. First characterized as a stroke by the ancient Greeks due to a sudden loss of consciousness – as if "struck down with violence" [6] – any rapid depletion of blood from the brain will likely prove fatal absent an immediate restoration of normal blood volumes. In the instance of heatstroke, clotting assumes the form of a blockage in the dissipation of heat from the skin's surface, with atmospheric humidity serving as the clotting agent.

The unique extremity of heat and humidity found in India is characteristic of a larger region stretching from the Arabian Peninsula across South Asia, where relatively shallow seas, such as the Persian Gulf and Red Sea, do not support as much cooling through deep water circulation as the Atlantic or Indian Oceans.

Unable to distribute absorbed solar energy as effectively through a deeper water column, these shallow seas give rise to very high rates of evaporation, fueling, in turn, the transport of unusually humid air across the subtropical deserts of the Middle East, Pakistan, and northern India. What results in the summer months is the most intense combination of extreme heat and humidity with densely populated settlement as found anywhere on Earth.

The implications of this combination of extreme heat and humidity for those unacclimated to these conditions – for example, to invading armies across history – can be quite severe. The first known record of a catastrophic heat-stroke event is provided by the Roman historian Dio Cassius, who recounts one of the most disastrous military campaigns of the Roman Empire, waged against a resisting population on the Arabian Peninsula in 24 BC. Entirely unprepared for the levels of heat and humidity commonly experienced in proximity to the Red Sea, more than half of the Roman army was lost to heatstroke before a single combatant was encountered. Cassius writes that "the malady proved to be unlike any of the common complaints, but attacked the head and caused it to become parched, killing forthwith most of those who were attacked" [7].

Heat stroke was a common occurrence through the period of the British occupation of India, with troops regularly succumbing to extreme heat expos-ures in the cramped and unventilated quarters of transport ships. Having failed to suppress an uprising in Calcutta (now Kolkata) in 1756, 146 British soldiers were imprisoned in a small, poorly ventilated cell overnight in summer condi-tions. Only 23 were alive the next morning [8].

And in more recent history, an estimated 20,000 Egyptian soldiers were lost to heatstroke during the Six-Day War with Israel in 1967, as the Egyptian army failed to supply sufficient water to its soldiers exposed to the summer heat and humidity of the Negev Desert [5].

In each of these instances, soldiers perished not only from an exposure to extreme heat but from a water imbalance – either too little in their bodies or too much in the atmosphere. What these historical narratives suggest in a world with declining freshwater supplies and rising water vapor is perhaps surprising: The greatest threat to our health from climate change is not a hurricane or a wildfire but an inability to sweat when our lives depend on it.

The Shrinking Map

One of the hottest cities in the world is also the coolest – at least along its sidewalks. With afternoon temperatures exceeding 100°F for more than four months a year, Doha – the capital city of Qatar, a small country in the Persian

Gulf region – is among the most extreme urban environments on Earth for a city with millions of residents. Compounding the heat is an average relative humidity in the summer months approaching 50 percent, resulting in maximum heat index values of about 140°F – more than sufficient to render a sidewalk stroll fatal. Despite these conditions, Doha's commercial districts are lined with high-end retail shops from around the world, complete with well-populated al fresco dining along the city's streets.

The solution to this enigma – how to provide a pleasant dining experience in an environment that exceeds physiologic thresholds for human heat stress – is to be found in football stadium design. In hosting the 2022 World Cup, Doha was confronted with the challenge of safeguarding the well-being of millions of visitors during a time of the year in which typical afternoon temperatures exceed the most intense heat wave conditions ever experienced by many traveling from outside the region. Doha adopted two strategies to address the heat threat. The first was to delay the World Cup by five months, to fall within the milder winter months of November and December. A similar strategy was adopted for the 2019 World Athletics Championships, also hosted by Doha, during which the women's marathon start time was shifted to midnight to lessen the intensity of the heat exposure (race time temperatures hovered around 90°F). More than 40 percent of the runners failed to cross the finish line, many requiring wheelchair assistance due to heat exhaustion [9].

The second strategy was to pump vast amounts of air conditioning into the football stadiums. This might sound like standard fare for large athletic facilities in hot environments, but the Doha World Cup planners had no intention of closing out the Sun and heat with a roof: The air-conditioned stadiums were open to the desert extremes. By positioning air conditioning vents immediately underneath each of the 40,000 seats in the Al Janoub Stadium – one of many newly constructed for the event – football fans could be cooled from their ankles up, before the chilled air, and the immense energy required to cool it, was lost to the desert winds. The same is true for sidewalk diners in Doha's commercial districts, where countless portable air conditioners are positioned table-side, allowing the customers and their butter pats to remain in a solid state (Figure 1.2).

Climate change is relentlessly characterized by the media, politicians, and, most regrettably, climate scientists as a temporal phenomenon – one unfolding over decades – but it is the changing spatial dimensions of current climate extremes that now most threaten us. The globally averaged environmental conditions projected for the middle or end of this century are no more extreme than those that we currently find in many parts of the planet today. If there was a tipping point beyond which the thermal extremes of cities in which millions

Figure 1.2 Use of air conditioning for outside dining in Doha, Qatar. Salwan Georges/Washington Post/Getty Images.

reside would become intolerable, we have passed it. For cities like Doha, it is only the resilience of the electrical grid that safeguards against the loss of tens of thousands of lives in a matter of hours – a collective ventilator to which every resident is now critically tethered.

While scientific terminology pertaining to life-threatening environmental conditions can be woefully understated, a recent addition to the technical lexicon bucks this trend. In the last few years, a growing number of technical papers, appearing in among the most staid journals of climate science, have included a new phrase in their titles and abstracts: *human survivability*. The use of this phrase is unusual in its seeming lack of precision, in that, given the wide range of climates in which humans have long settled – stretching from polar extremes to equatorial deserts – there are no historical ecological conditions considered outside the range of human adaptability. Yet this basic precondition for human settlement, at least with respect to heat, is no longer true.

If our geographic tolerance for environmental temperatures is surprisingly wide, our tolerable range for internal temperatures is surprisingly narrow. As any school-age child knows, just a few ticks above 98.6°F on the thermometer is sufficient to keep you off the school bus. The resulting condition – a fever – is widely defined by national public health agencies as a temperature of 100.4°F or higher, or, as more commonly measured, a temperature of 38°C. This number is remarkable both for its universality and for its circumscription. Travel the world over, from the polar-most outposts of Alaskan Inuit to the Tuareg peoples of Saharan Africa and you will find no measurable variation in the median core body temperature of healthy individuals. A fever in Moscow is measured as the same fever in Montevideo.

Beyond the constancy of this threshold, the human body can tolerate only the most minimal perturbation to its 98.6°F setpoint, and it is this immutable biological limitation that most starkly defines our present ecological circumstance. With only a single degree centigrade demarking the zone of well-being from that of illness, a shift in environmental temperatures that pushes us above this threshold constitutes a worrisome development. For the entirety of our impressive run at civilization building, commencing early in the epoch of the Holocene with the advent of advanced toolmaking, we have become quite skilled at adapting to an environment that has, with few exceptions, threatened to push our core body temperature *below* this critical threshold. We have managed environmental temperatures far colder than our setpoint temperature through technological adaptations in the form of clothing, insulated buildings, and fire-making. Indeed, it is precisely our adeptness in fire-making, in the form of hydrocarbon-driven industrialization, that is pushing environmental temperatures higher than our homeostatic temperatures.

Chief among the many challenges presented by a warming climate is this: Our tools for managing temperatures higher than our homeostatic setpoint are far more limited than those for managing temperatures lower than this threshold. Consider, for example, clothing. We have become very skilled over time in designing clothing for maximal insulation, varying both the materials and the thickness to achieve comfort in differing ranges of cold. Should a single layer of clothing prove insufficient in maintaining thermal homeostasis, we have the option of layering with additional clothing or with external insulators, such as blankets. Confronted with environmental temperatures that threaten to raise our core body temperatures, de-insulating with clothing for cooling is a much more limited adaptation than re-insulating for warmth, in that a single layer of clothing can only be removed once.

Likewise, no easily catalyzed chemical reaction for environmental cooling matches the simple efficacy of fire-building. Attainable with transportable ignition tools, plentiful natural fuel stocks, and universally available oxygen, fire is our species' most profound hedge against inhospitable environmental conditions. This is precisely the reason we have thrived in a world with environmental temperatures cooler than our homeostatic temperature in that our ability to regulate colder than optimal temperatures with fire has opened up a full planet to human migration and settlement. What equivalent tools can be transported in a backpack and combined with widely available fuel stocks to catalyze instantaneous environmental cooling in an uncomfortably hot setting? The absence of such tools, perhaps as the very product of our long-running adaptation to a cold planet, leaves us uniquely vulnerable in regions of the world in which environmental temperatures run higher than our homeostatic setpoint. And these regions are now expanding.

It is of course reasonable to observe that humans have lived for millennia in environments with temperatures periodically exceeding 98.6°F. What has changed, and only changed very recently, is our exposure to temperatures exceeding this homeostatic threshold in combination with humidity levels that inhibit our physiologic adaptations to environmental heat. While we often think of heat and humidity as positively related, with humidity levels seeming to rise during the summer months, the association between heat and humidity tends to exhibit a negative relationship, with relative humidity falling during the course of the day or course of the year, as temperatures rise. The reason for this is that relative humidity – the quantity of water vapor present in the air as a percentage of the maximum quantity of water vapor the air can carry – is directly related to the density of the air. To best grasp this association between atmospheric humidity and density, it is helpful to revisit the basic concept of temperature itself.

We associate temperature with the sensation of heat (or the absence of heat), but it can also be understood in terms of molecular motion. As the temperature of a gas or a liquid or a solid – let's imagine the handle of a stainless-steel saucepan for the moment – increases, the energetic motion of the elements that make up the saucepan increases in concert. In the case of our steel pan over a gas flame, the carbon and iron atoms present in a steel alloy will begin to vibrate with more intensity as heat is applied to the pan. This vibratory motion that starts in close proximity to the flame will be transferred along the pan to its handle as energized atoms agitate adjacent atoms in the solid material of the pan. It is the vibratory motion of the iron and carbon atoms in the saucepan handle that our hands experience as heat or as a rising temperature.

Were we to add water to the pan for boiling, the behavior of the water molecules would approximate that of the iron and carbon atoms but with one key exception: As the molecules of a liquid become energized with the addition of heat, they are free to move around within the full volume of the liquid, as opposed to vibrating in place. As the energized water molecules circulate within the pot, heat is gradually distributed throughout the body of water, as these energized water molecules displace cooler or less energized water molecules toward the heat source. This process of heat distribution via the circulation of energized molecules in a liquid or gas is referred to as *convection*.

The free movement of energized molecules in a liquid or gas via convection changes the liquid or gas in at least two respects. First, this convective movement distributes heat throughout the full volume of the liquid or gas over time, raising its temperature. Second, increasing molecular motion with rising temperature expands the volume of the liquid or gas, as repeated collisions between energized molecules create additional space between the molecules. When confined to the interior of a saucepan, the volume of water within the pan expands as it becomes heated, with a potential to overtop the pan itself. When confined to the far more expansive volume of the atmosphere, the available space between heated air molecules can increase significantly, resulting in a reduction in the density (the number of gas molecules per unit of volume) of the air.

Relative humidity will typically fall during the course of a summer day due to the decreasing density of the atmosphere with the addition of heat energy from the Sun. As the space between air molecules increases with rising temperature, the capacity of the air to carry additional water vapor also increases. If the water vapor content of the air remains fixed, the quantity of the water vapor as a percentage of the total water vapor the now lower-density air can absorb (i.e., relative humidity) will fall.

Of course, the quantity of water vapor in the air over the course of a warming summer day may also increase with a rise in evaporation from surface water and transpiration from trees and other plants. This increased evaporation, however, typically is not sufficient to keep pace with the falling density of air and its rising capacity to absorb water, and so relative humidity tends to fall over the course of a day, with lower levels observed in the late afternoon when temperatures are at their maximum.

Confronted with the need to carry out operations in extreme climates, the US military developed a new metric of combined heat and humidity in the 1950s to better forecast and characterize the risk of heat exposure. Referred to as "wet bulb globe temperature" the resulting metric is designed to more precisely pinpoint thresholds beyond which normal activities must be suspended. While indicators of combined temperature and humidity, such as the heat index, have long been measured and reported in weather forecasts, a key limitation of this metric is that it accounts for air temperature in the shade only. Shade-based measurements of temperature are desirable in that these allow for air temperature effects to be separated from *radiant* temperature effects. When we stand in the shade of a tree on a sunny afternoon, for example, we are warmed primarily by the temperature of the air. When we move away from shade, our skin is warmed both by the temperature of the air and by the direct receipt of radiation from the Sun.

In a world in which air temperatures rarely rise above our core body temperature 98.6°F, the combined effects of heat and humidity on our physiologic capacity to maintain homeostasis are of less import for human health than the threat of uncomfortably cold environmental temperatures, which can register 100 degrees or more below this homeostatic temperature in winter. On a warming planet, however, the hazard of extreme temperatures for human health is rapidly shifting toward the hot side of the comfort continuum – a condition for which we have fewer adaptive tools.

If your focus is on how the temperature and density of the atmosphere are influencing the operation of an aircraft, such as the length of runway that will be needed to achieve sufficient lift for takeoff, shade-based measurements of temperature are ideal, in that the receipt of direct solar radiation by the aircraft will not impact aerodynamic performance outside of its effects on air temperature. It is for this reason that virtually all airports measure air temperature in the shade, which is, by turn, the temperature reported in the local weather forecast – known technically as the *dry bulb temperature*.

If your focus is rather on human heat stress – particularly under conditions of direct solar exposure – it is critical that both air and radiant temperatures be measured, as both can raise core body temperatures. Wet bulb globe temperature measurements capture a more complete set of drivers of human heat stress by accounting for air temperature, radiant temperature, wind speed, and humidity, among other variables, that may not be routinely recorded at weather stations. Conceptually, wet bulb globe temperature assumes that a wet cloth has been wrapped around a thermometer and that the thermometer is then swung through the air in a circular motion and in full sunlight. This approach to wet bulb globe temperature measurement, which was the literal measurement technique prior to the development of electronic sensors, effectively measures what the air temperature would be if moisture was unlimited and evaporative cooling was maximized, which is the condition directly simulated by the incorporation of a wet cloth on a thermometer.

Wet bulb globe temperature provides a direct indicator of the temperature experienced by the human body mediated by our ability to release water to the surrounding air in the form of evaporated sweat. If relative humidity levels are high, little water will evaporate from the wet cloth around the thermometer, resulting in less cooling of the thermometer and a higher temperature. It is a metric that provides a single indicator of the environmental threshold beyond which humans cannot effectively cool themselves through perspiration, our body's most essential physiologic tool to maintain thermal homeostasis in hot weather. This threshold is generally recognized as 95°F wet bulb temperature, which is roughly equivalent to a dry bulb temperature of 110°F and a relative

humidity of 50 percent [10].[1] Due to the inverse relationship between temperature and humidity during the day, it is rare to observe a relative humidity of
50 percent at temperatures above 100°F, rendering a wet bulb temperature of
95°F quite extreme.

The availability of a single quantitative threshold to reliably predict health
outcomes in response to environmental conditions distinguishes extreme heat
from other climate-related hazards. For example, there is no single number on
the Saffir–Simpson Hurricane Wind Scale that is recognized as a definitive
threshold for human fatalities. Deaths during hurricanes are dependent on
a number of adaptive factors, ranging from the physical integrity of buildings
to the willingness of residents to seek shelter. Likewise, there is no accepted
wildfire class size, ranging from a relatively containable Class C fire to
a geographically expansive Class G fire, that is understood to with certainty
result in a loss of life. Prolonged exposures to wet bulb temperatures in excess
of 95°F are indicative of certain heat illness and likely death because there is no
adaptive measure one can take, beyond the availability of mechanical cooling,
to arrest the onset of core body temperatures outside of a physiologically
tolerable range. It is for this reason that a wet bulb temperature of 95°F is
now recognized, across national militaries and the wider realm of climate
science, as the threshold for human survivability [10].

Throughout the entirety of recorded human history, a combination of heat
and humidity sufficiently high to yield a wet bulb temperature of 95°F was
never measured until very recently and is unlikely to have occurred prior to the
advent of thermometers. The deadliest heat waves to date, including the 2003
European and 2010 Russian heat waves, which in combination resulted in more
than 100,000 deaths, both registered maximum wet bulb temperatures of about
83°F, with corresponding relative humidity values under 40 percent on the
hottest days in each event [11]. In the United States, the highest wet bulb
temperature values have been observed in the southeast and along the Eastern
Seaboard, and these most commonly top out at 85°F. Even the most recent
intense heat waves in hot and humid India and Pakistan have not crossed this
survivability threshold, despite reaching dry bulb temperatures in excess of
120°F [2].

---

[1] The survivability limit for humans is based on the "wet bulb" temperature, which is closely
related to the wet bulb globe temperature measure. Wet bulb temperatures are lower than wet
bulb globe temperature, as wet bulb temperatures do not account directly for radiant temperature,
which is measured with a black globe thermometer. In the interest of simplicity, I use these two
measures interchangeably. In practice, there is no consensus survivability threshold for wet bulb
globe temperature. Widely adopted guidelines for national militaries and athletic associations
generally designate 90°F wet bulb globe temperature as the safe limit for outdoor activities.

With temperature records falling regularly, and as we continue to venture into a climate unvisited by humans, many dates are significant. But some are more significant than others. In 1988, Dr. James Hansen, then the head of the National Aeronautics and Space Administration (NASA) Goddard Institute for Space Studies, would publicly and with a high degree of scientific certainty alert the US Congress to the mounting threat of a human-enhanced greenhouse effect, with his testimony and accompanying projections widely covered in the media. While the field of numerical climate modeling has continued to advance in the more than three decades since his testimony, Hansen's projections of how the planet would warm in response to continued and accelerating emissions of greenhouse gases were remarkably aligned with observations through the year 2020. As someone who turned eighteen the year of his testimony, this event carries particular weight, given that mine is the first generation to have knowledge of the unfolding ecological collapse throughout the entirely of its adult life – and the first to be unmoved by it.

The year 1998 would reveal for the first time the present-day potential for runaway climate change. While 1998 is no longer the hottest year ever measured, it represents the steepest year-over-year spike of the last 25 years, with a twelve-month increase in global temperatures of more than 30 percent relative to a long-term average and a two-year increase of more than 80 percent. The hottest year on record at the time of this writing is 2020 [12], but even the intervening twenty-two years separating these periods (1998–2020) would not produce an equivalent shift in global temperatures as experienced between 1996 and 1998, with the 2020 global temperature anomaly representing a 67 percent increase over 1998. The extremity of the heat that summer would spark massive wildfires not in arid California but in humid Florida, burning for three months and destroying more acreage than all but two of California's largest fires to date [13]. Worldwide, climate-related events in 1998 resulted in more than 32,000 fatalities and 300 million displaced from their homes. The estimated $89 billion in economic losses that year was 60 percent greater than the total economic losses from extreme weather events during the entirety of the 1980s [14]. From our vantage a quarter century beyond, the climate extremity of 1998 can be understood as the first impact of the iceberg upon the hull – the arcing flare of its mass and depth.

The year 2017 would bring not only the most destructive Atlantic hurricane season in global history but a statistical signpost as to how far down the climate change road we had progressed in the three decades since Hansen's testimony. With 17 named storms and 10 hurricanes, including Harvey, Irma, and Maria, the 2017 season resulted in more than 4,600 deaths in the US, almost $300 billion in property damages (exceeding the GDP of 171 countries), and

Figure 1.3 Interstate US-90 submerged in the aftermath of Hurricane Harvey, September 2017. Justin Sullivan/Getty Images.

widespread blackouts impacting more than 10 million residents, some of whom would lack power for almost a full year [15,16]. During Hurricane Harvey, a Category 4 storm, Houston, Texas would receive an excess of 60 inches of rain – more precipitation in twenty-four hours than typically falls in a full year (Figure 1.3). The quantity of rain deposited on Houston was found to be consistent with at least a 500-year storm – a flooding event so rare as to be expected to occur only once in the period between the discovery of the New World and today. Yet 500-year flood events in Houston are happening more frequently; in fact, 2017 was the third year in a row that Houston would experience a 500-year flooding event [17].

   This simple fact merits restatement: Houston endured for three years running a storm event so rare as to be expected to occur only once in twenty human generations. In the planetary epoch of the last 12,000 years – the Holocene – the probability of a 500-year storm occurring over three successive years would have been 1 in 125,000,000. In the current age of climate instability – increasingly referred to as the Anthropocene – the probability of such an occurrence is unknown, but it is inarguably higher. The lesson to be inferred from 2017 is not the need for new statistics but that, in an unstable climate, such statistics carry little predictive power.

To this brief list of climate change milestones, we must now add 2015 – the year in which the survivability threshold for environmental temperature was effectively reached. While the event went largely unnoticed at the time, a wet bulb temperature of 94.3°F was recorded in Bandar-e Mahshahr, Iran during an intense heat wave in which daytime dry bulb temperatures hovered close to 110°F and nighttime temperatures almost never fell below 90°F for a full week [18]. What most distinguished the Bandar-e Mahshahr heat wave, however, was the humidity. Situated astride the northernmost shores of the Persian Gulf, the city of 160,000 is warmed by winds from the shallow ocean with sea surface temperatures approaching that of a geothermal pool. For the entirety of a week, relative humidity values at times exceeded 80 percent and rarely dipped below 50 percent, even in the extreme heat of the afternoons. No data on heat injuries or deaths were released by the Iranian government.

The maximum survivability threshold of 95°F is, of course, the most optimal threshold – the threshold for a healthy, fit, well-nourished, and hydrated human. As described by one author, this limit "applies to a person out of the sun, in gale-force winds, doused with water, wearing no clothing, and not working" [10]. In other words, 95°F wet bulb temperature is what it takes to strike down the most resilient of human beings under conditions of no exertion, no insulation in the form of clothing, and in response to mediating environmental conditions that would never be encountered outside of a tropical storm. Given the unlikelihood of these conditions, the effective survivability threshold temperature is much lower than 95°F. According to the US military, 90°F wet bulb globe temperature is the threshold beyond which soldiers should immediately seek refuge and access to cooler conditions [19].

From this military threshold, we can adopt a rule for the rest of us. Given that wet bulb temperatures are neither commonly measured today nor reported, we need a set of guidelines for operating in a climate changed world, one in which, for the first time in human history, environmental temperatures can be sufficiently extreme to induce heatstroke in the most healthy of individuals – children, young adults, the fittest of athletes – in a matter of hours. While a wet bulb temperature of 95°F remains rare, in many parts of the world – including the United States, Europe, and the whole of Asia – exceedances of 90°F are occurring with increasing frequency.

So, here's a general rule: **> 100°F** with **50%** relative humidity. If the dry bulb temperature and relative humidity values forecast for the day meet or exceed these thresholds, seek shelter and cooling as you would seek protection from a hurricane. These numbers have real killing power; we should heed them.

With the effective attainment of the heat survivability threshold in 2015, and the continued uptick in global temperatures, the map for human settlement is at this very moment shrinking. While millions of humans reside in cities now too hot for safe habitation, the capacity to do so – or, at least, our collective capacity to overlook its cost in wellness and lives – is rapidly diminishing. Combined with the rising seas, an expanding zone of intolerable heat and humidity is already upending regional economic and political systems ill-prepared for rapid environmental change driven by unmitigated emissions of greenhouse gases. In some zones, adaptation is possible; in others, it is not. And so, the global human footprint that has been spreading since our initial forays from Africa more than a million years ago is, perhaps for the first time, losing its purchase on long-settled territory. No person alive today will be untouched by this retreat.

Changing the Weather

With a few seconds remaining in the game, the University Alabama men's basketball team was trailing Mississippi State University by two points. Having just been fouled by a Mississippi State player, the Alabama team had possession of the ball and, if it could convert the two foul shots, would send the game into overtime. Although no one inside the arena knew it at the time, the stakes for these two shots were much greater than a come-from-behind victory by Alabama and advancement to a championship game. With a strengthening tornado moving rapidly toward the Georgia Dome arena in Atlanta, where the game was drawing to its conclusion, a failure to convert either of the two penalty shots would release tens of thousands of fans to the exits, exposing them to a violent weather event for which no public warning had been issued.

Despite their rapid movement and powerful winds, tornados generally can be forecast well before touching ground. Referred to as a tornado "watch" in the United States, such vortical weather systems manifest first as intense thunderstorms and then with cyclonic rotating air masses that can be captured on weather radar systems, leading to a publicly broadcast tornado warning. Among the many ingredients needed for tornado formation is a high level of atmospheric moisture, which was not the case in the Atlanta area on the evening of March 14, 2008 – a key reason that no tornado watch was in effect (Figure 1.4). However, an additional ingredient – a strong updraft of heat over the urbanized region of Atlanta – was in plentiful supply and would help fuel the formation of a powerful tornado as the Alabama team was successfully completing its penalty shots – extending the game into overtime and keeping the fans in their seats, even as sections of the dome exterior were ripped away from the building [20,21].

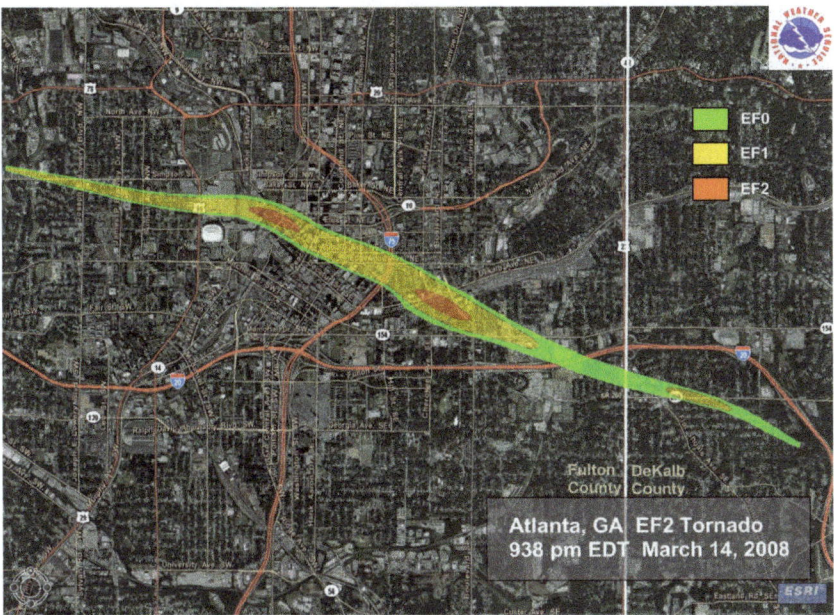

Figure 1.4 Path and intensity of a tornado in Atlanta, Georgia, March 14, 2008. US National Aeronautics and Space Administration, www.nasa.gov/topics/earth/features/atlanta_tornado.html.

It is often observed that no single extreme weather event can be attributed to climate change. Of all the technical caveats issued regularly by climate scientists, this is perhaps the most puzzling. All extreme weather events today are influenced to some degree by climate change, with a growing number attributable in full to higher baseline temperatures, higher baseline sea levels, and, in many regions, lower baseline levels of soil moisture. Given that sea levels are increasing every year, across the extent of the global oceans, a hurricane of historical intensity will generate a higher storm surge than in the past due to higher baseline water levels; all hurricanes moving forward are rendered more destructive as a product of higher sea levels. The same is true for the intensity, scope, and duration of heat waves.

Both global sea levels and temperatures are higher today than in the past due to human enhancement of the global greenhouse effect. Analogous to the glass panes of a greenhouse, gases naturally present in the Earth's atmosphere, such as carbon dioxide, absorb longwave radiation emitted from the Earth and are warmed by this radiation, elevating atmospheric temperatures higher than would be the case without these gases. The Moon, for example, has a much less dense atmosphere than the Earth (due to its low gravitational field) and lacks any measurable greenhouse effect, yielding temperatures too extreme for life. The Earth's natural greenhouse effect, in concert with the presence of liquid water and oxygen, is part of our ecological life-support system – we would not be here without it.

Human activity has not given rise to the global greenhouse effect, but our emissions increase the concentration of greenhouse gases naturally present in the atmosphere and, by turn, enhance planetary temperatures. This has long been true. During the period of the late Roman Empire, for example, the extensive burning of plant matter for fuel and metalworking released significant quantities of methane into the atmosphere, a potent greenhouse gas, likely contributing for a brief period to modestly higher global temperatures [22].

Since the dawning of the Industrial Revolution, both the quantities of greenhouse gases emitted from the burning of coal and the impact of these emissions on the planetary greenhouse effect have driven a clear and continuing rise in average global temperatures that reaches to the present moment. The retention of additional solar energy at the planetary level fuels more extreme weather through numerous drivers: the atmosphere carries more water for rainfall and the oceans, once warmed by the atmosphere, steadily rise through thermal expansion and polar melting. As described famously by Dr. James Hansen in his 1988 testimony to the US Congress, it is these human-enhanced emissions of greenhouse gases that have pushed the Earth outside of its long-established pattern of climate stability, resulting in a greater probability each year of weather events exceeding historical norms. Human activities have, as he described it, loaded the climate dice, with each roll now more likely to produce an outcome that was once rare.

This well-recounted climate change narrative has been extensively tested with both present observations and historical data – it is today, quite likely, the most exhaustively validated theory in the realm of the physical sciences. First proposed in the 1890s and supported with evidence every year since, the theory of a human-enhanced greenhouse effect has not been challenged in a scientifically meaningful way for almost half a century.

Importantly, however, a human-enhanced greenhouse effect does not explain the full extent of warming at the level of regions or, in particular, the level of cities. At physical scales generally less spatially expansive than that of continents, changes in the composition of the land surface can also drive temperatures higher than would occur in the absence of human activities. As noted in the Prologue, one such change is deforestation for agriculture or cattle grazing. As the extensive conversion of forest to grasslands reduces the quantity of rainfall that can be captured by forest ecosystems and retained as soil moisture, the resulting grassland ecosystems tend to be drier than the forest ecosystems that precede them, contributing to less evaporation from soils and plants, less rainfall, and, as a direct result, higher temperatures. This enhancement of temperatures as a product of regional land use change can serve to further amplify warming brought about through the emissions of greenhouse gases at the planetary scale, but the two warming mechanisms are physically distinct – one can occur without the other.

In cities, the conversion of forested areas, crops, or wetlands into buildings, streets, and parking lots also produces a warming effect. Referred to as the "urban heat island effect," late afternoon temperatures in cities are measured to be anywhere from 2°F to 20°F higher than proximate rural areas, yielding a magnitude of warming that is typically greater than that brought about through the global greenhouse effect [23]. The urban heat island effect is driven by four specific changes in cities. First, the removal of forest cover and other green plants reduces both shading and the rate at which water is returned to the atmosphere via evaporation and transpiration, processes that cool the urban environment akin to the cooling of skin through perspiration. Much like a forest converted to a grassland, the resulting degradation in photosynthetic activity and vegetative biomass from land clearance will bring about in most regions of the planet a warming effect, even prior to the development of the city itself.

Second, the introduction of impervious construction materials to the cleared landscape, including concrete, asphalt, and roofing materials, greatly enhances the "thermal capacity" of the resulting built environment, or the quantity of heat energy that can be stored by these materials and released back to the atmosphere, elevating temperatures. The extensive use of paving and roofing materials in cities also serves to waterproof large swaths of the urban environment, limiting the infiltration of rainwater into urban soils and reducing average soil moisture. The process of urbanization, in this respect, is effectively a process of desertification, yielding a climate much drier and hotter than that displaced by the footprint of the city.

Having transformed the landscape into a more efficient machine for the absorption, storage, and reemission of heat energy than the natural land covers that are displaced, the largest cities further compound their warming effect through the creation of large urban canyons. Constructed through the clustering of tall buildings or the introduction of walled roadways below grade (or both), the prevalence of vertical surfaces in cities, fashioned from stone, concrete, and glass, traps outgoing longwave radiation much like the radiative trapping of greenhouse gases. Impeded from escaping the urban canyon, some fraction of outgoing radiant energy is reabsorbed and reemitted by building walls, further elevating their temperature and enhancing the urban heat island effect.

A final mechanism through which cities elevate their own temperatures results from the extensive use of energy – most of it derived from fossil fuels – to generate electricity and power mechanical processes. Governed by the second law of thermodynamics, some fraction of the energy used to carry out mechanical or electrical work – whether in the form of a vehicle engine, an industrial press, or an illuminated light bulb – will be transformed into waste heat energy and emitted from tailpipes, smokestacks, or radiated directly into the ambient air. This waste heat energy is sufficiently concentrated in cities to elevate air temperatures and further enhance the urban heat island effect. In this sense, cities are heated by solar energy from two eras: present-day sunlight that is more efficiently captured and stored by urban materials than by the natural landscape and ancient sunlight, captured by green plants eons ago and embalmed as fossil sources of energy. It is this compounding of solar energy – the present augmented with the ancient – that renders large cities among the most extreme environments found today on Earth.

In combination, these four drivers of the urban heat island effect – the removal of natural land covers, the introduction of materials with a high thermal capacity, the creation of urban canyons, and the release of copious quantities of waste heat energy – not only change the climate of urbanized regions over time; these characteristics of cities change the weather. If climate is the long-term pattern of weather experienced across a region and weather events are short-term, day-to-day meteorological phenomena, cities have been clearly demonstrated to have a pronounced effect on both temporal scales – ranging from the thirty-year average temperature to the intensity of a multiday heat wave to the instantaneous wind speeds of a tornado. This unique characteristic of urban areas – a tendency to amplify the solar receipt of energy – is rapidly accelerating the pace of climate change in cities.

Occupying less than 2 percent of the global land surface but home to more than half of the global population, cities today are simultaneously the most climatically hazardous and commonly chosen pattern of human settlement. Residents of cities experience the same intensity of greenhouse-induced warming as found outside of urban areas and compound these heat exposures with the even greater thermal liability of the urban heat island effect. Data collected over a fifty-year period in and outside of the largest cities across the United States find the majority to be warming at about double the rate of the planet as a whole, with the urban heat island effect responsible for more than half of this warming year-over-year [24]. As a result, every observable dimension of heat waves in cities – including the frequency, intensity, duration, and seasonality – is increasing and, in most instances, increasing rapidly [25]. To live in the center of a large US city today is to confront a risk of heat morality 200 percent greater than for those living on the urban fringe, where heat islands are less pronounced [26]. If a human-enhanced global greenhouse effect has loaded the climate dice, in cities we are playing the game with more than two dice – and still throwing for doubles.

But cities, to a much greater extent than rural areas, can alter their climate odds. Rendered by decades of development more absorbent of solar energy, less capable of cooling through an exchange of moisture between the land surface and atmosphere, and far more generative of their own fossil heat emissions, cities can be physically redesigned to moderate these warming effects and, in some instances, exhibit temperatures lower than their predevelopment landscapes. Cities can and should be places of less heat and carbon-intensive modes of travel than nonurban settings. Cities can be modified to capture and store more rainwater than is lost through extensive surface sealing, recharging soil moisture, bolstering drinking water supplies, and leveraging water stores for direct cooling when climatically advantageous. Cities can be made once again, as they were prior to their industrial remaking, responsive to their local climates, and adherent to their ecologically delimited carrying capacity. Cities can, in short, shape their climate fate.

In the remainder of this chapter, I will consider four specific classes of urban design strategies – referred to collectively as *urban heat management* – available to cities to significantly moderate both the pace and the extremity of urban warming. The design and materials-based strategies considered range from the ancient to the experimental and are not uniformly suitable for every urban setting. While these strategies provide a physical toolkit for moderating the extremities of heat in cities, they address only one dimension of a broader approach to managing heat risk. Equally important are emergency response protocols that can be activated during periods of extreme heat – such as early warning systems for heat waves and the provision of cooling centers – as well as the enhancement of individual adaptive capacity, including well-insulated housing and access to transportation for evacuation, when warranted. While these latter topics are not the focus of the following discussion, they remain critical elements of climate resilience plans and should be implemented in concert with physical changes to the built environment of cities.

While each of the strategies to be considered has been implemented in some combination and at some scale by the most forward-looking municipal governments, no city to date, in any reach of the planet, has undertaken the physical transformation of its built environment needed to safeguard its population from the coming heat. With the typical municipal budget allocation for heat management still falling short of monies appropriated for the annual patching of potholes, no city to date has embarked upon the almost complete resurfacing of the built environment – including every road, every parking lot, and most every roofing surface – that will need to be programmed in the present decade. If the resulting greening and lightening and watering of the urban environment is ancient in its conception, it must be radical in its implementation to maintain the viability of cities in a climate changed world. Nothing less will meet the moment.

Greening

July 21, 2019, was a hot day in Cambridge, Massachusetts – the hottest of the summer. With typical July temperatures in this small city within the larger Boston area reaching to only 85°F, temperatures exceeding 100°F – as reached on that day – are rare, and these extremes are not experienced in all neighborhoods. East Cambridge is a neighborhood characterized by dense, low-rise housing, narrow streets with limited tree cover, and much less park space than other Cambridge neighborhoods. With little tree cover available for the shading of streets or cooling through transpiration, East Cambridge is often more than 10°F hotter than other areas of the city, and it is also the neighborhood in which the fewest residents have access to air conditioning. In light of the elevated heat risk confronted by residents, a colleague and I were invited by the city to carry out a sort of do-over for July 21. If we go back in time and remake East Cambridge to be more heat adaptive, to what extent could extreme summer temperatures be lowered? To what extent could hospital visits and heat-related deaths be reduced?

The do-over for July 21, and every day during the summer of 2019, would take the form of a climate model simulation. Despite the grave risk posed by extreme temperatures in cities today, few cities invest much effort in monitoring this risk. The truth is we have no direct means of knowing what the temperature was in East Cambridge on July 21, 2019. The most proximate permanent weather station to this neighborhood is at Boston's Logan Airport, a few miles to the east and situated at the mouth of Boston Harbor – a very different climatic setting than a dense residential neighborhood only partially bounded by water. While there are security cameras positioned at numerous locations across Cambridge, by comparison there are very few temperature sensors for which historical data are available, and none of these is located in the neighborhood confronting the greatest heat risk. Driven by extensive information on the physical characteristics of every city block in Cambridge – every building, rooftop, roadway, greenspace, and tree – combined with regional weather patterns for every hour of the day, urban scale climate models provide a highly reliable tool for simulating weather where we lack direct observations. Such climate models allow us not only to travel in time, reproducing historical weather conditions, but to simulate how the weather on a particular city block would change in response to alterations in the immediate physical environment, such as the addition of trees or reflective roofing materials.

Prior studies focused on how a greening of cities could lessen their heat risk have shown a great potential for a restoration of natural land covers, where annual rainfall is sufficient, to substantially cool the urban environment. In what now numbers in the thousands of technical studies – some dating to the 1960s and supported with data from cities situated across every populated continent – the literature on *urban heat management*, a subfield of urban climatology focused on cooling cities through a more climate-responsive physical design, is definitive: green plants in urban environments lessen the extremity of heat, often dramatically [27]. This lessening of heat scales with the horizontal extent and vertical density of green cover, carries no significant cooling penalty in the winter months, and is highly compatible with other aims of climate change adaptation, such as flood control. The cost of increasing and maintaining green cover in cities is not insignificant but is more than offset by the avoided energy costs associated with lower summer temperatures [28]. And the greening of private property enhances its value, yielding an economic return to landowners, as well as to city governments in the form of increased property taxes [29]. Among all of the grim headlines associated with climate change, this simple observation tacks differently: *We can make our cities more resilient by making them more beautiful.*

The extent of cooling provided by tree canopy has been found to be relatively consistent across different climate types. Our heat assessment work in Cambridge and other cities finds that the shading of 1 percent of a neighborhood's area with tree canopy lessens late afternoon temperatures by about 0.2°F. This rate of cooling has been found to be roughly consistent across cities located in cool and humid settings, such as Cambridge, and those located in hot and arid conditions and is drawn from studies focused on cities with an average tree canopy cover of between 10 percent and 40 percent [28,30,31,32]. For reference, the average tree canopy cover for a sample of large US cities was found to be around 30 percent [33]. In these cities, as a general rule of thumb, an increase in neighborhood-wide tree canopy of 10 percent, if well distributed, can be expected to lower temperatures by about 2°F. This is a substantial level of cooling and, in many large cities, is on par with the level of additional warming projected to occur by mid-century, enabling some or all of this anticipated warming to be offset [31,32]. But cities should aim much higher than for a 10 percent bump in tree canopy.

What extent of citywide tree canopy is possible? This question is of less direct import than the relative canopy cover of residential areas and zones with significant pedestrian activity. In these zones, a minimum canopy coverage of 40 percent has been observed to yield the greatest increase in cooling per tree added [34]. But this is a minimum – residential districts of many large cities would be well served by a canopy coverage of between 40 percent and 60 percent, representing in many North American cities a doubling of average citywide tree canopy.

To understand if such an extent of tree canopy is feasible, we must only look to contemporary examples. Returning to Cambridge, a dense city by US standards, about 30 percent of the city is presently overlaid with tree canopy. According to the US Forest Service, if we account for areas available for planting, including non-road impervious areas, such as parking lots, 65 percent of Cambridge could be overlaid with tree canopy [35]. Applying the same Forest Service methodology, a similar canopy potential exists in New York City (64 percent), with 50 percent of Manhattan found to be capable of supporting tree canopy based on present-day development patterns [36]. Similar numbers are found across the Eastern United States – 67 percent of Washington, DC is plantable [37]; 69 percent of Philadelphia [38]; 71 percent of Baltimore [39]. While somewhat less green than eastern cities, many cities of the western US can support tree canopies in excess of 40 percent, such as Boise (57 percent) [40], Portland (52 percent) [41], and semi-arid Sacramento (45 percent) [42]. The vast majority of the US population lives in cities where a much more extensive canopy of tree cover is feasible with no change in present-day development patterns and allowing for future population growth (Figure 1.5).

A substantial increase in tree canopy is not delimited by suitable land or rainfall patterns in most cities, but it is delimited by an insufficient allocation of resources. How much would it cost to increase the tree canopy from 30 percent to 50 percent in a dense city like Cambridge?

An analysis of thirty-four studies drawn from around the world (with the majority in North America) finds the average annual cost of urban tree planting and maintenance to be about $38 per tree [43]. More recent work focused on the annual cost and maintenance of street tree planting and maintenance in California cities finds this cost per tree to be about $110 [44]. As the planting of trees along streets and in proximity to other types of impervious cover carries the greatest cooling potential, a rounded estimate of $100 per tree per year – closer to the cost tallied for street trees in California cities – provides a reasonable basis to understand

Figure 1.5 Street tree canopy in Brooklyn Heights, New York. Ellen Isaacs/Alamy Stock Photo.

the budgetary implications of increasing the average citywide tree canopy in most regions of the planet. Based on our work in Cambridge, an additional 60,000 trees would be needed to attain an average canopy cover of 50 percent in every neighborhood. Assuming a conservative annual cost of $100 per tree (as not all newly planted trees are assumed to be street trees), the additional annual cost to Cambridge of undertaking this tree planting and maintenance program would be $6 million, representing less than 1 percent of the current annual budget, about 4 percent of the annual budget for policing services, or an approximate doubling of the current budget allocation for urban forestry [45].

In outlining these cost comparisons, my intent is not to trivialize an additional annual expense of $6 million to a modest-sized US city. It is to highlight that the costs of expanding the urban forest, even in the most dense of urban environments, are consistent with other annual expenditures on critical urban services and infrastructure. The costs of expanded tree planting and maintenance in Cambridge represent less than a quarter of the annual cost of servicing bonds issued for the construction and maintenance of sewers [45]. Cambridge need not conclude that trees are more important than storm sewers in a climate changed world, but the city's leaders must recognize, perhaps for the first time, that urban heat management is equal in importance to public welfare as flood management.

An additional point to make on the flooding versus heat management comparison is that we have no technological analog to engineered sewer systems for lessening heat exposure in cities. As urban populations confront for the first time an intensity of heat that will not permit even the most healthy of individuals to work a full day outside, or typical summer temperatures that render little league baseball unsafe, the development of a technology capable of measurably cooling cities through a partial blocking of solar radiation and a phase change in water – and one that could do so at a cost representing less than 1 percent of the annual city budget – would be considered a miracle of innovation. Were an engineered solution available for ambient cooling in cities – such as the sidewalk air conditioning of Doha – it would undoubtedly require a tremendous investment of material and energy and would carry a significant carbon footprint. Tree planting, by contrast, requires no research and development, produces no ambient noise (or waste heat emissions), and would have a net effect of reducing rather than increasing atmospheric carbon dioxide. What we have in the form of trees is a climate management tool surpassing the performance of any that humans could hope to create – and one that, amazingly, would soon dominate the landscape again in most cities if we simply stopped suppressing their growth.

The hottest day of the Cambridge summer in 2019 would have been remarkably less hot were at least 50 percent of every neighborhood in the city overlaid with tree canopy. In East Cambridge, the neighborhood found to exhibit the highest late afternoon temperatures across the summer, the high temperature that day would have been 6°F cooler – sufficient to more than offset the projected magnitude of warming by mid-century and possibly end of century. For the summer as a whole, the number of hospital visits for heat-related illness in East Cambridge would have been about 25 percent lower, and the number of heat-related deaths almost 50 percent lower. Through tree planting alone, Cambridge can effectively hold the line on heat for another generation – an outcome that no greenhouse gas emissions control technology or program, no matter how aggressive, could feasibly achieve.

If there is anything radical about the proposal to increase urban tree canopy in large cities to 40 percent or 50 percent or even 60 percent, and to do so in the most densely developed areas, it is only in the required expansion of our conventional thinking about infrastructure. Cities allocate tens of millions of dollars every year to the construction and maintenance of sewer systems and road systems and water delivery systems because these systems are fundamental to the viability of urban life. In a climate changed world, the expansion and maintenance of dense tree canopy are no less fundamental to the viability of cities. If we can put rovers on Mars, we can put trees in dense street corridors, throughout parking lots, and even atop buildings (Figure 1.6). This is not

Figure 1.6 Trees incorporated into building rooftop and terraces, Vancouver, Canada.
JSM Images/Alamy Stock Photo.

a technical feat beyond our capacity or our budget, but it cannot be achieved with a mindset that urban trees are principally ornamentation.

Reaching these canopy targets will require the engineering of planting spaces to support healthy root growth, the monitoring of all public trees for regular maintenance of health, and a well-calibrated nursery program to rapidly replace trees when lost to extreme weather, pests, and other causes of canopy loss [46].[2] None of these requirements render an expansive urban forest less feasible than other types of urban infrastructure, and none would place undue stress on municipal budgets. For slowing the rise of temperatures in all but the most arid cities, trees are the most basic of a new class of adaptation infrastructure – the surest arrow in the resilience quiver.

Shading

With urban tree canopies in decline in most large cities of North America [33], one land use category is experiencing a quiet revolution in adaptation planning for heat: the playground. The focus of now long-standing public

[2] The New York City Parks Department, for example, has created an online map displaying the location of its 692,892 trees planted across the city, including the maintenance history for each as well as a running tally of climate benefits (see [46]).

health campaigns undertaken by the US Centers for Disease Control and Prevention and the Trust for Public Land, a nongovernmental organization focused on greenspace development and design, a significant number of playgrounds have increased tree canopy or installed shade structures to lessen solar exposure among children [47,48,49]. Initially designed as a response to a rising epidemic of skin cancer, artificial shade structures in playgrounds and other athletic facilities are the leading edge of a trend toward the greater shading of urban environments as an adaptive response to rising temperatures. While less effective in lowering ambient temperatures than tree canopy, shade structures nonetheless yield substantial cooling benefits and can be deployed over spaces inhospitable to tree canopy. A long-standing feature of urban design in hot and arid climates of the Middle East and Mediterranean basin, urban shade structures are being deployed over entire street corridors to reduce the duration of intolerable heat exposures in commercial areas. The extensive integration of natural and artificial shade in large cities stands as a promising but largely untested adaptation strategy in other regions of the world.

The effectiveness of shade structures in lessening human heat stress is well established by hundreds of studies spanning all major climates of the globe. Importantly, the provision of shade by such structures – including covered transit stops, pergolas over walkways, mesh screens, fabric canopies, and a wide range of materials, both modern and ancient in construction – is found to significantly reduce heat stress without substantially reducing air temperatures. Shade structures moderate heat stress through the attenuation or complete blockage of incoming solar radiation, which often has the effect of lowering radiant temperatures without significantly lowering air temperatures. As noted, the physiologic effect of radiant temperature is experienced when moving from direct sun to complete shade on a hot day. Given that environmental heat is transferred to the human body via the direct receipt of solar radiation and via convection from the air that surrounds our bodies, the moderation of either source of heat energy will lessen heat stress. While the blockage of radiant energy by shade structures does have the effect of lowering the temperature of the air under the structure, the relatively small fraction of the urban environment shaded by such structures does not yield much of a cooling effect on the extensive volume of the urban airshed. But these structures do cool our bodies measurably while we remain within the shade they cast.

Here, again, the extent to which dry bulb air temperature – the environmental variable most commonly reported by weather stations – fails to directly account for the effects of other environmental variables on heat stress, including the direct receipt of solar radiation and humidity, militates for the use of more direct measures of heat stress, such as wet bulb globe temperatures. To understand the potential role of shade structures in lessening human heat stress, a study in hot and arid Phoenix, Arizona measured a full set of heat exposure variables in the shade of trees and artificial shade structures within an urban park. While the maximum dry bulb temperature on a summer afternoon reached 104°F, a more complete index of heat stress found temperatures to be equivalent to 124°F in full sun. Moving into the shade of trees within the park was found to lower the heat stress index by more than 20 degrees, to 101°F – still hot but less life-threatening in the arid climate of Phoenix. The cooling effect of shade structures was found to be about 80 percent that of tree canopy, lowering the heat stress index to about 105°F [50].

Both the extent to which shade can lessen heat stress and the relative cooling effectiveness of natural and artificial shade have been found to be largely consistent in studies sited in desert, subtropical, and tropical climates. Studies focused on pedestrian zones in Colombo, Sri Lanka, Nagoya, Japan, and Rome, Italy, for example, find the provision of shade from adjacent buildings or other structures to lessen a heat stress index by between 20 percent and 30 percent – a bit higher than the range found in Phoenix [51,52,53]. Studies comparing the relative effectiveness of natural and artificial shading find shade structures to yield at least 80 percent of the cooling effect of tree canopy [50,51,54], demonstrating the utility of artificial shade for managing heat stress in lieu of or in concert with tree canopy.

For cities confronting a rise in heat exposures beyond the range of human tolerance, the development of extensive and interconnecting shade corridors will be required to sustain pedestrian activity and support critical infrastructure for emissions reduction programs, such as the wider use of transit systems (Figure 1.7). Fabricated shade structures, in particular, provide a climate management strategy supportive of moderating the impacts of climate change in cities while simultaneously lessening carbon emissions – a set of strategies I refer to as *adaptive mitigation* [23]. Already deployed in some cities as scaffolding for solar photovoltaic (PV) systems positioned over parking lots, the integration of shade structures with micro-grids – decentralized renewable energy systems capable of supplying power to the grid when centralized power stations and transmissions systems are disabled by extreme weather or other causes – is a low-hanging strategy that enhances resilience while lowering greenhouse emissions. The development of shade structure systems to the extent required in many cities will simultaneously create a new layer of urban surfaces ideally positioned for the deployment of solar PV

Figure 1.7 Shade structure over street in Granada, Spain. Benedek/iStock.

collection cells across dense urban corridors, precisely where the demand for renewable energy is greatest.

Fabricated shade structures provide an essential heat adaptation strategy in highly trafficked zones of cities situated in climates too arid to support extensive tree canopy. As urban populations are exposed to extreme temperatures not only during episodic heat wave periods but increasingly throughout long stretches of the summer months, heavily shaded, "cool corridors" will need to be fashioned from both natural and fabricated shade structures across entire urban districts. In cities with rainfall patterns supportive of tree growth, the integration of fabricated shade structures with trees may provide opportunities for enhancing the resilience of urban trees to storm damage from the physical tethering of trunks to permanent shade structures. A long-practiced defense against wind damage in hurricane-prone regions, trees in fruit orchards are often stabilized with support systems including guying straps or wires and root ball anchors [55]. The use of permanent shade structures as tree support systems, in concert with the selection of tree species most resilient to wind and drought, provides an opportunity to both lengthen the reach of such corridors across urban districts and extend the life of valuable green infrastructure.

Reflecting

As we ascend ever further up the climate change mountain, a proposal to slow the rate of planetary warming through geoengineering – a set of Hail Mary strategies to cool the planet through untested, global-scale efforts to, for example, inject aerosols into the upper atmosphere or seed the global oceans with nutrients to increase the rate at which carbon dioxide is absorbed – is being actively debated by climate scientists. The risk of these approaches is not only in their unknown secondary effects (like an unanticipated disruption in global agriculture) but also in the necessary scale at which such approaches must be implemented. Experimentation with planetary physics risks irreversible changes, but the same is not true for urban-scale geoengineering. Beyond the restoration of natural land cover, such as tree canopy, cities can also be physically modified to reflect away incoming solar radiation, producing a localized cooling effect. A technique long practiced in premodern white-washed villages of the Mediterranean, where the extensive use of white paint and naturally reflective building materials lessen the quantity of energy absorbed from the Sun, new classes of engineered materials are capable of cooling buildings to such an extent that air conditioning may not be needed except on the hottest days.

Enhancing the reflectivity of surface materials cools the air by short-circuiting the usual pathway through which sunlight warms the Earth. With a limited number of exceptions, the communities of plants and animals that comprise the Earth's biosphere have evolved to absorb solar energy – often as much as can be gathered across a small surface area. The dark green hue of an oak tree leaf, as one common example, assists the tree in retaining much of the solar energy that falls upon it, driving, in turn, photosynthesis and the set of biological functions carried out by the living tissue of the tree. Lighter-hued materials, by contrast, such as fresh snow, reflect away more incident solar radiation than is absorbed, a property that slows the rate at which snow will melt on a clear day. Evidence of this strong reflective property of snow is manifest in the sunburned faces of unprepared skiers, an outcome of the upward flux of solar energy reflected from the snowpack.

The reflection of sunlight away from the Earth's surface cools the air by lessening the quantity of heat energy absorbed and reemitted from the ground. Governed by the same properties that drive the global greenhouse effect, reflected solar radiation is permitted to pass back through the atmosphere, without warming the air, because this radiation has not been transformed into longwave, infrared radiation – the form of radiation absorbed by greenhouse gases. Shortwave radiation (mostly visible light) is transformed into longwave

radiation (mostly infrared, thermal radiation) when absorbed by surface materials and reemitted at a lower temperature and (by physical law) a *longer*
wavelength than the visible radiation received from the Sun. Like a mirror
that reflects your image back at you, highly reflective surface materials simply
return shortwave, solar radiation to the atmosphere without converting it into
longwave radiation.

Knowledge of these physical properties of radiant energy enables
a reengineering of cities to be less absorbent of solar energy through the use
of more reflective building materials in new construction or the application of
reflective coatings to existing buildings. Measured on a scale from 0 to 1 and
referred to in technical parlance as "albedo," the reflectivity of cities can be
increased through the use of lighter-hued roofing and paving materials, such as
a white surface coating on the flat roof of an industrial building.[3] To assess the
extent to which high albedo roofing and paving materials could lower temperatures on a hot summer afternoon, we modeled such a scenario for the same
recent summer period in Cambridge, Massachusetts for which we measured the
benefits of planting more trees. In doing so, we assumed a greater increase in
the reflectivity of rooftops than for surface paving in the form of streets and
parking lots, as the use of high albedo materials at ground level can create glare.
While a high level of solar reflection at the level of rooftops is less of a concern
for visibility, too much reflection at the street level can be hazardous for drivers
and cyclists. With this in mind, we assumed all of the streets of Cambridge were
converted from darkly hued asphalt paving to more lightly hued concrete and
that all rooftops were converted to levels associated with commercially available "cool" roofing materials, depending on whether the building had a flat
(more reflective) or pitched (less reflective) roof.

Our modeled "do-over" for the hottest day of the summer in 2019 finds that
reengineering the city to be more reflective would lower temperatures by
between about 0.5°F to 1°F in the late afternoon, depending on the neighborhood. This variability in cooling benefits is largely driven by the surface area
available in each neighborhood for cool materials – in higher density neighborhoods with more roofing area, the benefits of reflective roofing are greater.
Overall, each 10 percentage point increase in albedo was associated with about

---

[3] It should be noted that conventional "cool materials" used for roofing and paving applications are
also engineered to exhibit a high "emissivity," in concert with a high albedo. Thermal *emissivity*
is a property of materials measuring the rate at which absorbed solar energy is returned to the
atmosphere. Materials with a high emissivity (also measured on a scale from 0 to 1) are able to
release heat energy back to the atmosphere efficiently and quickly, which limits the quantity of
heat energy retained by the material. By storing greater quantities of energy received from the
Sun, lower emissivity materials will experience a greater rise in temperature than higher
emissivity materials, assuming all other thermal properties are held constant.

0.5°F of cooling at the neighborhood level – suggesting that a 20 percent increase in roofing and paving reflectivity, an obtainable goal in Cambridge, could lessen late afternoon summer temperatures by about 1°F. This rate of cooling compares well with numerous other studies focused on cool materials, which tend to range from roughly 0.5°F to 1°F of cooling with each 10 percent increase in surface albedo and average about 0.7°F [56]. Based on our modeling work in Cambridge and other cities, a greater use of cool materials for roofing and paving materials is roughly 25–50 percent as effective as tree canopy in lowering neighborhood-scale temperatures during the summer, assuming the same land area is converted to either treatment.

A second thermal property of urban construction that can be engineered for greater cooling is the capacity for phase changes to moderate peak temperatures. Similar to the effects of a phase change in water-to-water vapor, a phase change in building materials from a solid to a liquid phase can alter the timing at which solar energy absorbed by a building is returned to the atmosphere in the form of heat. The encasement of paraffin (wax-based) capsules in the voids of exterior walls, for example, enables a greater percentage of the energy absorbed during the day to be stored by the building wall, as the solid paraffin converts to a liquid through melting, without contributing to higher temperatures. At night, once the air temperature cools, the stored heat energy will be slowly released as the paraffin material undergoes a reverse phase change back into a solid. The use of phase change materials in roofing tiles has been found to outperform high albedo materials by about 30 percent, suggesting the potential for a combination of high albedo and phase change materials to lessen air temperatures more significantly than conventional cool materials alone [57].

Recent advances in super-cool coatings for building roofs suggest the potential for cool roofing materials not only to reduce the quantity of heat emitted by urban structures but to measurably cool the surrounding ambient air [58]. Derived from the pigmentation of white paint with barium sulfate nanoparticles, the resulting "ultra-white" coating has very high levels of material albedo (0.98) and emissivity (0.96), serving to reflect away almost all intercepted solar energy. While conventional cool materials lessen the quantity of heat returned to the atmosphere relative to traditional roofing materials, such super-cool materials can exhibit a surface temperature that is actually lower than that of the overlying atmosphere, serving to reduce the temperature of the air much like an outdoor air conditioning system and transmitting no thermal energy into the building itself. The demonstrated cooling effect is so great that the use of these materials for single-story structures could eliminate the need for mechanical air conditioning inside the buildings on most days, easily offsetting the additional cost of incorporating these coatings into roofing materials.

Importantly, the potential for this new class of ultra-white cool roofing materials to remain highly reflective over time in response to weathering is not yet demonstrated. Nonetheless, a growing emphasis on such adaptive mitigation strategies for cities – those designed to reduce the drivers (greenhouse gas emissions) and impacts (extreme heat) of climate change simultaneously – is yielding scientific advances that could substantially cool cities through a scale of geoengineering far less likely to carry irreversible side effects than planetary-scale interventions. If combined with other heat management strategies better suited to pedestrian environments, such as tree planting, it is not beyond our reach to substantially lower temperatures in cities and to offset, in turn, many decades of warming brought about through the global greenhouse effect.

Watering

First measured in nineteenth-century London, the long-lived and growing intensity of the urban heat island effect – perhaps the first scientifically documented evidence of human-driven climate change – may begin to abate in some of the world's largest cities over the coming decades. In what would represent the first substantial federal investment in climate change adaptation for heat in the United States, the Infrastructure Investment and Jobs Act of 2021 allocated hundreds of millions of dollars for the expansion of green infrastructure and cool materials in urban areas. As the performance of thermally engineered building materials continues to advance, coupled with an expansion of tree canopy and shade structures in response to more extreme heat events, urban heat management carries the potential to render cities cooler than surrounding rural areas. The potential for such "urban cool islands" is growing as freshwater supplies are declining in agricultural areas around cities. Outside of monsoon periods in India, for example, a significant number of cities exhibit lower temperatures during the day than do nearby rural areas characterized by extreme drought conditions and little vegetation [59]. A related outcome in agricultural areas confronting dwindling groundwater supplies is the likelihood of more intense heat waves with a reduction in regional rates of cropland irrigation [60].

A growing extremity of temperatures in rural areas with a shift toward drier conditions attests to the power of a single environmental variable in regulating air temperatures: water. If the release of water by green plants through transpiration is a dominant driver of air temperatures, to what extent could the irrigation of urban greenspace and streets offset temperatures during heat wave conditions? To what extent would urban irrigation elevate heat risk through the enhancement of humidity and wet bulb temperatures? While there is not today an extensive scientific literature focused on these questions, the available evidence suggests that urban irrigation – where water supplies can support it – can measurably lower heat exposures in cities, even when accounting for a modest rise in humidity. A recent review of studies assessing the potential for enhanced irrigation to lower afternoon temperatures in cities finds modeled day and nighttime temperature reductions to range from about 1°F to 10°F, with no study finding evidence of an increase in temperatures [61]. The greatest potential for reducing temperatures through a more extensive watering of urban vegetation is found in hot and dry cities, such as Phoenix or Las Vegas, where humidity levels are typically low. A common finding of these studies is that increasing levels of urban irrigation (or water availability) are associated with decreasing cooling benefits, suggesting that this strategy carries more limited benefits for cities in humid climates.

To be viable in arid cities, where freshwater supplies are already stressed, reclaimed sources of non-potable water would be needed. In cities constructing *sustainable urban drainage systems* (SUDS), a set of techniques through which stormwater and greywater (recycled domestic wastewater free of fecal contamination) are captured and stored for reuse, the irrigation of urban vegetation may provide a suitable use of these reclaimed water supplies. To assess the potential benefits of urban irrigation for heat management in Adelaide, Australia – a city with expanding greywater infrastructure – a recent study modeled the effects of the irrigation of urban greenspace (with no increase in the total area of urban vegetation) on temperatures during heat wave conditions. The results found average daytime temperatures fell by more than 4°F during an intense heat wave, with maximum temperatures falling by more than 15°F in zones with high soil moisture levels, suggesting a strong potential to deploy watering as an emergency response strategy during heat wave conditions. A heat index, accounting for the combined effects of temperature and humidity, was also found to fall in response to widespread urban irrigation during heat wave conditions – during both the day and the night [62].

A decade after what remains among the deadliest weather disasters on record – the European heat wave of 2003 – French researchers carried out experiments focused on street watering during conditions of extreme heat in the cities of Paris and Lyon. In each experiment, surface and air temperatures were

Figure 1.8 Street watering during heat wave in Moscow, Russia, July 2021.
Alexander Sayganov/SOPA Images/LightRocket via Getty Images.

measured across adjacent portions of the same urban street, with one block
sprayed by watering trucks with greywater at regular intervals and the adjacent
block left untreated. An earlier study used a climate model to estimate the
benefits of street watering citywide during heat wave conditions. The results
demonstrated a modest cooling effect at the scale of the single street corridor,
with a maximum air temperature reduction of about 1.5°F, and a potentially
greater effect from citywide street watering, estimated at up to 3.5°F [63].
Reflecting the need to account for both the temperature and the humidity effects
of street watering, wet bulb temperature effects were measured during a similar
experiment in Lyon, finding a maximum reduction of about 1°F in this most
direct indicator of heat stress [64].

Performed in cities with high levels of summer humidity, these experimental
studies in Paris and Lyon, combined with modeling studies focused on greenspace irrigation in hot and arid climates, support the deployment of urban
irrigation as a strategy to moderate temperatures during heat wave conditions
(Figure 1.8). While the cooling effects estimated from irrigation do not match
those achievable through enhancements in tree canopy, a key advantage of
irrigation is the potential to deploy it as an emergency response strategy during

dangerously hot weather with existing resources, such as watering trucks, and with no needed change to the built environment of cities. While the development of greywater collection and storage systems is needed to enable such an approach in arid regions, many large cities confronting growing water scarcity are already investing in integrated water management systems for a wide range of water needs.

The demonstrated cooling benefits of the various heat management strategies considered in this chapter suggest the potential to combine techniques to maximize temperature reductions during summer conditions. To date, a large number of modeling studies has been carried out focused on the combined effects of urban vegetation and cool materials. The median modeled cooling benefits of these now commonly adopted strategies, across a diversity of countries and climate types, is found to be about 3.5°F – a level of cooling associated with a wide range of assumptions pertaining to the area of new green cover or the level of achieved surface reflectivity. A much more limited number of studies has focused on the use of all four strategies – greening, shading, reflecting, and watering – and finds a combined effect of these approaches on maximum daily temperatures to be in excess of 6° F in arid climates, a level of cooling well exceeding the performance of any single strategy and sufficient to mostly or fully offset the urban island effect [65].

In short, an extensive greening of cities combined with managed soil moisture and the use of thermally responsive building materials carries the potential not only to offset continued warming in urban areas brought about through the global greenhouse effect, estimated at present to be about 0.6°F per decade,[4] but to arrest further warming while accommodating additional population growth. No foreseeable greenhouse emissions control program, of any geographic scale – ranging from the individual city to the entirety of global human settlement – can come close to attaining this level of urban cooling in the present century. The body of evidence on urban heat management is scientifically robust, widely validated, and unambiguous on this basic point: Cities can slow and quite feasibly reverse their present rate of warming through built environment strategies implemented at the urban scale and falling fully within the legal authority of most municipal or provincial governments. Urban heat management can be adapted to a wide range of regional climate conditions, requires a budgetary allocation comparable to other types of critical urban infrastructure, and is highly compatible with

---

[4] The average decadal rate of change in mean global temperatures since 2000 is estimated from the NASA global mean temperature anomaly dataset as 0.57°F (https://data.giss.nasa.gov/gistemp).

other climate policy goals, such as the reduction of greenhouse gas emissions. Urban heat management can render our cities more resilient and more livable at the same time.

This full-throated endorsement notwithstanding, a radical redesigning of cities to lessen the intensity of heat is not the solution to climate change. Adapting to rising temperatures at the small geographic scale of cities is not the same as changing the underlying global physics driving the global greenhouse effect. The ability to slow the rate of warming in cities is at best a mode of triage to buy much-needed time for global emissions controls to stanch our carbon bleed. And these approaches will not be sufficient to keep cities in the most extreme settings within the bounds of human habitability. The need for sufficient water supplies – fresh or reclaimed – to sustain some level of green cover in urban environments will not be possible in all regions confronting ever-deepening water scarcity.

Radical Adaptation for Heat

It has been my purpose in this chapter to highlight the ways in which extreme heat – a rising threat to human health in cities worldwide – is distinct from other modes of environmental management. Cities cannot be walled off from rising levels of heat in a manner similar to the construction of levees for rising volumes of water, nor can cities be drained of heat from the construction of a series of underground pipes. Heat cannot be trucked away to a landfill or purified at a centralized treatment facility. To manage heat most effectively, cities will need to move beyond traditional management approaches to a set of strategies that are much less centralized and much less engineered. Amplified by the design of the city in all locations and during every hour of the day, heat can only be moderated through a set of approaches that are equally diffuse spatially and temporally. Every building, every roadway, and every open space will need to be assessed for its potential to support a canopy for shading, whether vegetative or structured, for its potential to be rendered more reflective, and for its potential to retain moisture.

In a word, our approaches to managing heat in cities must be much more *dispersive* than conventional modes of environmental management. The first of four principles of radical adaptation to be considered, a dispersive approach to climate adaptation represents a sharp departure from large, centralized public works facilities of the nineteenth and twentieth centuries and recognizes that the physical structure of the city itself must be the focus of our management efforts, as opposed to its byproducts alone. Inherent in this shift from a centralized to a dispersed set of management strategies is a shift, at least in part, from publicly owned facilities to privately owned properties. Through a radical approach to climate adaptation, the city must be redesigned across its entire extent to minimize the hazardous effects of heat. The same is true of water.

2

Water (Too Much)

Not all of the registered names are memorable. *Full of Walgreen's Trash*, for example, or *Love Your Neighbor* hardly reveal the wellspring of creativity one expects to find in the wards of New Orleans. Others are a bit businesslike in their geographic specificity: *To the Right of Lee's House* or *Renee's Childhood Home*. The really good ones, however, show some flair for the task: *Pump Up the Basin*, *Kim Catchbasinger*, and – my personal favorite – *Catcher in the Right of Way*. In the fall of 2018, New Orleans would undertake a public works project unlike any other: The city would assign responsibility for clearing debris from its street stormwater catch basins to nearby residents. In exchange, the residents were invited to name the catch basin on their block. While the program would fail to attract much national attention, it represents a significant milestone nonetheless, as it marked the first time a major US city would acknowledge its inability to safeguard its population from the devastations of climate change.

Perhaps no other element of physical infrastructure signifies our current moment more than the lowly catch basin. Largely invisible to daily life, outside of the occasional parallel parking exercise or lost ball, storm drains are the gateway to a subterranean shadow city that makes urban life possible. The first urban storm sewer systems were little more than reinforced creek banks – a course of bricks to stabilize a natural waterway as the volumes of rainwater running off from the surrounding city streets intensified. But soon the streets themselves would need to run atop the streams, and the ancient watercourses – many having given the early city its initial form – were diverted into tunnels and buried. With an ever-increasing proportion of the city's land area rendered impermeable to rainwater, such human-made water conveyances ultimately would be needed under every street, giving rise to a hydrologic life support system for the urban population above.

For residents of New Orleans, the value of this life support system was discovered anew in August 2017. More than a decade after the complete inundation of Hurricane Katrina, expanded storm surge defenses at the city's periphery provided a higher level of flood protection than ever achieved on US soil. The more than $14 billion invested in the city's levee system and a new set of storm surge gates installed at major canal outfalls into Lake Pontchartrain have provided sufficient confidence to city residents and businesses that the rising waters around the city can, for the time being, be kept out [1]. What was not fully considered, however, was the success of these flood defenses in keeping the water in.

The topography of New Orleans is often described as a bowl but perhaps a better description is a bathtub situated in a large and rising body of water. Water can enter the tub by overtopping the sides, as occurred during Hurricane Katrina, or as rainfall from the sky. When received as rainfall, the tub's drain and accompanying pumps are designed to remove the water – but only if the drain plug has been removed. As with all city stormwater systems, the thousands of catch basins along the streets of New Orleans are the drain in the bathtub and thus must be regularly cleared of debris to allow the passage of stormwater. In advance of a heavy storm on August 5, 2017, the levees held and the pumping stations were (mostly) operational, but the drain plug remained stubbornly in place.

The rain event that day would qualify as a 50-to-100-year storm – an intensity of rainfall expected to occur one or two times in 100 years, based on historical rainfall patterns in New Orleans. While meteorologists are quick to note that it is possible for more than a single 100-year storm to occur in a particular location in the same 100-year stretch, it is not what we expect to happen based on the local history of storms. But happen it does, and with much greater frequency in a climate changed world. Rather than the 50 to 100 years expected to pass between such events, the time that had elapsed between the August 5 storm and the most recent event of similar intensity was approximately two weeks: On July 22, a 50-to-100-year storm had deposited almost 5 inches of rain on the city in a single hour, flooding streets, vehicles, and buildings [2].

In both of these instances, the levees held and the pumping systems were largely operational. The problems encountered were less an instance of systems failure than one of systems design. A large number of catch basins were clogged with trash and other debris, greatly slowing the rate at which stormwater could be removed from streets. In the months that followed, New Orleans would undertake a $7 million effort to remove debris from these catch basins and enlist residents in ongoing maintenance. The chief bounty of these

Figure 2.1   Street debris near a catch basin during Mardi Gras celebrations in New
Orleans, Louisiana. David Jennings/Alamy Stock Photo.

excavations was more than 46 tons of Mardi Gras beads, a massive hangover of
plastic that, with each passing of the second line parades, had accumulated over
years, rendering the extensive upgrades in pumping capacity since Hurricane
Katrina far less effective at removing storm water (Figure 2.1) [3].

A second problem encountered pertains to the overwhelming volume of
water that must be removed from a tub-like city when more than 9 inches of
rain fall over the course of three hours. With the present pumping system in
place, the result of more than $2 billion invested in pumping capacity over
a decade, only a single inch of rainfall can be removed in the first hour
following the commencement of rain, falling to one-half inch for each hour
thereafter. This is not the attenuated performance of a system clogged with
plastic beads but the rate of drainage the system was designed to achieve with
well-maintained catch basins and fully operational pumps. As summed up by
the superintendent of the local water board after the flood, "[i]f you are asking
me to drain 9 inches of rain, I need six times the pumping capacity, six times the
drainage pumps and six times the canals. I don't need three or four more pumps,
I need 400 or 500 more" [4]. In New Orleans, both the frequency and the
intensity of rain experienced in a climate changed world simply exceed any

flood management system that the city, state, and federal governments are willing to construct.

The most pressing threat to New Orleans today is not the next storm surge from the Gulf of Mexico but the next slow-moving rain event – an event that no levee system of any height or length can keep at bay. Like so many coastal cities today, New Orleans cannot simply wall itself off from the rapid environmental changes underway in southern Louisiana. If the Big Easy is to persist in a world of once-in-a-generation storms that now recur with regular frequency, what the city most needs within its levee walls is not less water but more.

Flooding the Swamp

Two years after the New Orleans heavy rainfall events, another leveed city of the United States would flood in response to more than 4 inches of heavy downpour in a single hour. Almost 100 miles from the nearest coastline, Washington DC is often overlooked as one of the leading global cities at risk to flooding from climate change, but it ranks among the most imperiled cities in the United States. Situated at the confluence of two rivers impacted by sea level rise, and constructed atop hundreds of acres of filled wetlands, the US capital city is the lowest lying of any major national capital city in the world, with the next lowest lying among the Group of Twenty (G20) nations perched several feet higher than Washington DC [5]. A torrential rainfall event in July 2019 would swell local streams by more than 10 feet, inundate the National Mall and monuments, and, not for the first time, flood the basement of the White House. It is estimated that by the end of the present decade, less than eight years hence as I write, the capital of the world's self-proclaimed sole superpower will have flood waters lapping at the steps of its monuments two to three times a week [6].

The selection of a site for the national capital of the nascent United States in 1790 is widely characterized by historians and in contemporaneous writings on the debate as a compromise between northern and southern states. While there was a general consensus that a location near the Mason–Dixon Line – an established border between land grants of the British Crown to resolve a boundary dispute generally taken to divide the southern and northern colonies – would be optimal for the nation's capital city, the site at the confluence of the Potomac and Anacostia Rivers fell well below this line, thereafter recasting agrarian and slave-holding Maryland as a northern state. Of equal interest in site selection to George Washington, the first US president and a leading proponent of the site, was not only the conjoining of northern and southern colonies but a direct water connection to the emerging western states via the Ohio River. To realize this vision, Washington would lobby the US Congress to undertake the construction of a series of westward-reaching canals, culminating in the Chesapeake and Ohio (C&O) Canal system, completed in 1831. Perhaps incidentally, President Washington owned a Potomac River shipping company that would profit considerably from the creation of a federal canal system along its banks [7].

The outcome of these many political considerations would be a site for the federal city championed by its boosters as the perfect compromise between north and south, east and west. As with many societal decisions of the late Holocene – the geological epoch that commences with glacial retreat about 12,000 years ago and is argued to conclude with the emergence of a human-driven climate signal in the twentieth century [8] – that the very geographic advantages of the site selected for the US Capitol building might one day require its relocation in no way occurred to its founders. Herein lies among the most daunting aspects of an unstable climate, for it not only complicates societal choices moving forward but compromises those made long ago. As we transition rapidly from a stable to an unstable climate, the most pressing consideration for the viability of the US capital city is no longer cartesian but topographic.

As water was pumped from the White House basement in July 2019, it was observed by members of the press corps that the long-standing commitment of politicians to "drain the swamp" of DC corruption was at long last being realized. But the swamp waters seem destined to return. As the levels of the global oceans steadily rise, some coastlines are experiencing higher daily tides than others. The mid- and southern Atlantic coastlines are found to be experiencing some of the most rapid rates of sea level rise of any coastline worldwide. With global mean sea levels projected to experience an approximate 0.2-meter increase by 2050, the mid-Atlantic coastline, at some locations, will exceed 0.4 meters [9]. And this pace of sea level rise impacts not only the coastlines but the volume of water of all rivers, bays, and wetlands that are influenced by the daily tidal fluctuations. Despite its distance from the Atlantic Ocean, the Potomac River experiences a tidal swing of as much as a meter every day as it flows along the banks of Washington DC, enough to create nuisance flooding in low-lying areas of the city on an increasingly frequent basis (Figure 2.2).

The massive annual emissions of greenhouse gases driving atmospheric temperatures higher year after year are also impacting the global oceans. With roughly 70 percent of the planetary surface occupied by seawater, rising heat in the atmosphere is transferred to the upper layers of the global oceans and, more gradually, distributed to deeper layers of water through oceanic circulation. The implications of rising temperatures for the planet's oceans

Figure 2.2 Tidal basin flood in Washington DC. Bill Chizek/iStock.

are many, with the most direct effect manifested in a thermal expansion of seawater. By the same principles governing the rising volume of water in a pot placed over a flame, seawater will experience an increase in kinetic energy with the addition of heat from the atmosphere, serving to intensify molecular motion and lowering the density of the water. As the seawater grows less dense, it expands, serving to raise the volume of the oceans without the addition of water from land or air.

Compounding this process of thermal expansion, of course, is the addition of massive amounts of water from melting ice sheets atop Greenland and Antarctica and from glaciers distributed around the planet. The accumulation of precipitation received over millennia, the ice presently melting atop Greenland is estimated to have been deposited as snowfall more than 40,000 years ago. At the base of the Greenland ice sheet lies snowfall deposited more than a million years ago, well before the emergence of a species of ape that would accelerate its return to water.[1] The present rate of planetary ice loss is estimated to be 1.2 trillion tons a year – 60 percent greater than twenty years ago [10].

---

[1] Importantly, the melting of floating icebergs and sea ice does not contribute to sea level rise, as the mass of this ice already influences the volume of the ocean. The expansion of this melting iceberg water with warming, however, does drive sea level rise.

For much of the industrial era, the principal driver of sea level rise has been the thermal expansion of seawater. The most recent science, however, suggests that the scale has now tipped toward accelerating ice melt upon land, a shift that carries weighty implications for global weather. After a long period of contributing only about 5 percent of the total annual meltwater to rising seas, the Greenland ice sheet is more recently estimated to have experienced a more than fivefold increase in annual ice loss, accounting now for more than 25 percent of the annual rise in global sea levels [11]. As the lowest latitude (and elevation) major reservoir for land ice, Greenland is experiencing higher summer temperatures than Antarctica, and, importantly, the temperature of the ocean waters surrounding Greenland are higher, accelerating the melting of glaciers where they reach the coast [12].

In total, the annual contribution of melting ice sheets and glaciers to rising seas is now more than double the contribution of thermal expansion [11], and this shift is further compounding not only the rate of sea level rise but the rate of planetary warming. Much like the urban heat island effect described in Chapter 1, the loss of highly reflective ice on land or in the Arctic Ocean has the effect of lowering surface albedo. The newly exposed land following ice loss or the return of dark seawater to the Arctic Ocean reduces the quantity of solar radiation reflected away, increasing in turn the quantity of heat absorbed and returned to the atmosphere to amplify the global greenhouse effect.

The most recent Intergovernmental Panel on Climate Change (IPCC) Assessment Report, a compendium of the consensus science of climate change issued every five to seven years, estimates that the global oceans are rising at an average rate of 3.7 mm per year. This annual rate is increasing rapidly and is found to be almost 200 percent higher than during the mid-twentieth century [9]. Accounting for this rate of acceleration, the IPCC projects a mean global sea level rise of 0.2 meters (0.65 feet) by 2050, rising to more than 0.8 meters (2.6 feet) by 2100 under our present emissions trajectory [9]. The rate of sea level rise in proximity to many cities, however, will be considerably higher due to regional variability. Washington DC, for example, is projected to see the Potomac River rise by 0.4 meters (1.3 feet) by 2050, 100 percent greater than the global mean projection and higher than any other major city on the East Coast [13].

For a city already experiencing regular nuisance flooding from the Potomac River at high tide, an almost doubling in the extent of sea level rise over the present decade (0.10 meters over baseline levels in 2020 vs. 0.19 meters in 2030) will inundate significant portions of southwest Washington DC during storm events and contribute to the weekly frequency of nuisance flooding projected to occur by 2030. This decadal rate of sea level rise exceeds that of any other city on the Eastern Seaboard of the United States, including New York, Charleston, and Miami, and is only surpassed by much smaller cities on the Gulf of Mexico, such as Galveston, Texas, which is expected to experience an increase in sea level of 0.11 meters in a single decade (from 0.14 meters to 0.25 meters) [13].

The rate of sea level rise in US cities of the East and Gulf Coasts surpasses that forecast for any other region of the planet. Venice, Italy – a city wholly constructed on wood piers in a bay – exhibits a present rate of sea level rise that is only about 50 percent of that of Washington DC and New York. No coastal city in Western Europe is experiencing a present rate of sea level rise much higher than the global average of 3.7 mm per year. And virtually no coastal city in all of Asia, Africa, or South America exceeds present rates of sea level rise in the most threatened coastal cities of the United States, with only Manila, Philippines surpassing Washington DC but falling short of the rate of Galveston, Texas.[2]

Compounding the effects of sea level rise in many coastal cities is a concurrent process of land subsidence. Large coastal cities are sinking year-over-year due to several phenomena, including soil compression from urban development and – in the mid-Atlantic region of the United States and else-where – a long-term geologic process associated with glacial retreat. Even today, glacial and other land ice remains the largest reservoir of fresh water on the planet, accounting for more than 60 percent of the global supply. The sheer mass of this accumulated water in the form of snow and ice measurably compresses the land surface as glaciers advance, serving in turn to create a bulging effect in the soil substrate farther afield as a product of displacement (a process referred to in technical parlance as "glacial isostatic adjustment"). With the retreat of most glaciers from the upper reaches of North America at the start of the Holocene, the attenuation of glacial mass continues to slowly reverse this bulging effect in the land substrate across parts of the mid-Atlantic United States, including the Washington DC area [15]. As a result, Washington DC is sinking by a rate of about 2 mm per year, while the Potomac River rises by a rate of about 6 mm per year – yielding a "relative" rate of sea level rise of about 8 mm per year [13][3] – more than double the global average of 3.7 mm per year.

---

[2] These data on sea level rise are drawn from the NASA Sea Level Rise Projection Tool, a global dataset constructed from observations and projections included in the IPCC's Sixth Assessment Report. A small number of cities not included in this dataset, such as Jakarta and Semarang in Indonesia, are found elsewhere to be experiencing a higher rate of relative sea level rise than US cities, as a product of land subsidence [14].

[3] All rates of sea level rise reported in this chapter are relative rates, reflecting the effects of both sea level rise and land subsidence.

Figure 2.3 Roadway erosion in New Orleans driven by land subsidence. William Morgan/Alamy Stock Photo.

In coastal cities already attempting to manage more frequent flooding, the installation of pumping systems to remove flood waters is perhaps the least costly strategy in the short term, but regular pumping is likely to exacerbate flooding problems over time. Returning to New Orleans, the reliance on a network of large pumping stations and canals to remove rainwater has the unintended effect of depleting urban soils of moisture between rain events. This regular removal of rainwater, combined with the extensive use of impervious surfaces found in all cities, desiccates urban soils, creating a compression effect over time. Evidence of this regular soil compression can be seen in the city's notoriously crumbling roadways – an unwinnable battle of asphalt maintenance in a sinking city (Figure 2.3). At present, parts of New Orleans are subsiding at a rate as high as 50 mm (about 2 inches) per year, yielding a relative rate of sea level rise that is almost four times greater than other parts of southern Louisiana and is second globally only to Semarang, Indonesia, where aggressive pumping of groundwater is lowering the city's elevation by as much as 120 mm (4.7 inches) per year [14].

Urban subsidence and rising seas would have the effect of intensifying flood events in coastal cities absent an increase in the frequency or intensity of rain events. That the volume of stormwater received per rain event is simultaneously increasing across most cities of the planet – both coastal and inland – attests to the growing challenge confronted by cities to manage ever-greater volumes of stormwater and explains why some of the most extensive flood damage experienced to date has occurred far from any coastline. Intensifying rainfall worldwide is a product of the same physical properties governing the expanding seas: An increase in the temperature of the atmosphere lowers its density, increasing the capacity of the atmosphere to absorb and transport water vapor. As rising global temperatures also have the effect of evaporating ever-greater quantities of water from the oceans, inland water bodies, and soils, both the availability of water vapor and the capacity of the atmosphere to carry it are increasing with planetary warming.

The rate at which the atmosphere absorbs greater quantities of water vapor with rising temperature is so constant across space and time it has been christened with the names of the scientists who first measured it in the nineteenth century – Rudolf Clausius and Émile Clapeyron – and is a fundamental component of the science of thermodynamics. The "Clausius–Clapeyron" rate tells us that with each 1°C rise in atmospheric temperatures (1.8°F), the quantity of water vapor absorbed by the atmosphere will increase by about 7 percent. With global temperatures now standing at 1.2°C (2.2°F) higher than preindustrial levels, a quantity of water equal in volume to more than half of the rivers on Earth has been transferred to the atmosphere. By mid-century, a volume of water exceeding the total found in rivers planetwide will have been transferred to the atmosphere, further seeding unprecedented storm events [16,17].

The breaching of this growing atmospheric reservoir during precipitation events is yielding more intense rain and flooding episodes, even in regions of the planet becoming more arid with rising temperatures. Analyses of rainfall data from the last several decades, a period in which warming accelerated, show a significant increase in both the frequency and the intensity of extreme rain events in all regions of the planet [18,19,20,21]. With continued warming, the annual number of extreme rain events is projected to approximately double with each 1°C (1.8°F) rise in global temperatures [21].

The combination of enhanced evaporation and more intense rain events with rising temperatures yields an unexpected outcome in many regions of the planet, where local climates are becoming drier and more flood-prone at the same time. As long-term droughts unfold in many regions of the world, including northern China, southern Europe, the western United States, and almost the entirety of India, these same regions have simultaneously experienced record-breaking rainfall and flooding events. As perhaps the leading global example of this phenomenon, Chennai – a megacity of more than 10 million inhabitants in southeastern India – would be the second major global city, following São Paulo in 2015, to effectively exhaust its drinking water reservoirs under drought conditions, in 2019, requiring a massive convoy of water trucks and trains to sustain the urban population. Two years later, Chennai would receive more rainfall in November 2021 than in almost any other month in its history, but this temporary deluge would only provide short-lived relief from what appears to be a long-term shift toward a drier climate [22].

The linkage between persistent drought conditions and extreme rain events can be found in the thermal engine that drives both phenomena. As higher temperatures increase evaporation, moisture that is depleted from the land surface is often returned through a smaller number of more extreme storm events. Both historical observations of rainfall and climate models find a reduction in the frequency of what were once typical rainfall events, while the number and intensity of rare events is rising [20,21]. In short, the light drizzling of a weekly spring shower is being replaced by the deluge of an autumn tropical storm, a pattern for which urban stormwater systems are ill-equipped to manage. The concentration in time of precipitation events results in deeper depletion of soil moisture and surface water between events, desiccating the land surface and rendering it less permeable to rainfall once it returns. Unable to absorb and store extreme volumes of rainfall in single events, most stormwater is simply lost to surface runoff, enhancing the potential for flooding and providing only short-lived relief from drought conditions.

Worldwide, more than 70 percent of the global land surface is projected to experience greater aridification over the present century, ranging from tropical rainforest climate zones to deserts. At the same time, a shift toward more extreme precipitation patterns is expected for both humid and arid regions, with flood intensities projected to increase in these climate types by more than 70 percent and 60 percent, respectively [16]. The result is a need for every city, in every region of the world, to plan for a growing intensity of both flood *and* drought conditions. Doing so will require not an expansion of our pumping systems or a raising of our levees but a radical rethinking of what a city should be. Long designed to repel water, the climate-adaptive city must catch it; long designed to displace its most natural areas, the climate-adaptive city must restore them. In the ill-conceived war between the city and nature, nature has won. There is no path forward that fails to acknowledge this simple truth.

A River Runs Through It

At first glance, the image of Tan Hoa, a small settlement perched near the Kien Giang River in Vietnam, captures a now familiar scene: a rural village flooded

Figure 2.4 Floating homes in Tan Hoa, Vietnam. Oxalis Adventure, www.oxali
sadventure.com.

during seasonal monsoon rains of unprecedented historical intensity. Red clay
tile rooftops extend from the brown flood waters, alongside the uppermost
branches of otherwise submerged trees. In October 2020, Vietnam would
experience some of the most intense flooding in its history, with six tropical
storms or cyclones washing over regions of the country and depositing as much
as 20 inches of rain in a single day [23]. In what has become a familiar though
no less tragic litany, hundreds would drown in the flooding; hundreds of
thousands would be displaced; and hundreds of thousands of cattle and other
farm animals would be swept away [24]. What distinguishes the image of
a flooded Tan Hoa from other nearby villages is the handful of green-sided
structures tethered to poles and floating atop the flood water, front doors
opening onto the roofline of the adjacent house (Figure 2.4).

   The seed for the floating Tan Hoa houses would take root in post-Katrina
New Orleans. Appalled by the tens of thousands of houses that were lost to the
floodwaters and the slow pace of what continues today as a multigenerational
rebuilding process, Dr. Elizabeth English, then a young professor at Louisiana
State University, had an audacious idea: Thousands of New Orleans shotgun
houses, creole cottages, and bungalows – the historic building stock of middle-
and low-income neighborhoods situated at the lowest elevations – could be
retrofitted to rise and fall with future floodwaters. So sure was she of the idea

that a grant was soon secured to test the idea in a house-sized water tank. The demonstration project was definitive: The installation of inexpensive buoyant materials under the ground floor of an existing house, such as foam blocks, combined with the erection of mooring poles to which the structure can remain tethered above its foundation could safeguard modest-sized homes from all but the most extreme flooding conditions, avoiding post-storm displacement and the substantial expenditures required for rebuilding.

The concept of an amphibious house is not new. The widespread use of houses built atop rafts not intended for water transport, in contrast to a houseboat, dates at least to temporary logging communities established in the Pacific Northwest in the early twentieth century. A more recent iteration of the idea is rapidly growing in the Netherlands, where entire neighborhoods of floating but permanently moored houses have been under construction since the early 2000s. Such communities are designed not only to weather the influx of water in an unstable climate but to capitalize on the extensive surface area of water bodies across a country largely reclaimed from ancient peat and marsh-lands. For the Dutch, amphibious housing has become synonymous with affordable housing, and it is precisely this feature of a home liberated from land that illuminates the brilliance of English's idea for houses intended to float only in emergencies. Amphibious housing is a bulwark against retreat and an adaptation radical in its potential to embed resilience into the most climate-vulnerable communities not last but first.

The cost of a buoyant retrofit for a modest-sized, one-story house is estimated to be less than $4,000 for the most basic system and can be installed by two workers and without the need for heavy machinery [25]. For new homes in flood-prone areas, the use of a buoyant foundation adds an estimated 5–10 percent to the cost of construction – a modest expense when compared to that of rebuilding the home after a flood event or permanent elevation of a structure, a strategy increasingly adopted in communities subject to recurring inundation. Beyond the cost of elevating homes on tall piers, this approach often creates a vertical separation between neighboring houses and subjects the structure to much greater risk of wind damage. Despite the clear economic, social, and environ-mental advantages of buoyant retrofitting for lower-income communities at risk of devastating flood damage, floating homes presently are not eligible for flood insurance in the United States, where the Federal Emergency Management Agency (FEMA) asserts the approach is insufficiently tested, more than a decade since its technical demonstration. Lacking such governmental impedi-ments, the strategy is gaining ground in other flood-prone regions of the planet, such as Vietnam.

The greater wisdom to be grasped from the idea of floating houses is that, as the Dutch put it, we must make "room for the river." Unable to bend ecological forces to our will, we must adapt our cities not to prevent flooding but to weather it. Through a national plan adopted in 2006, the Netherlands initiated a program to expand the flood zone of three major rivers that were breaching conventional flood management structures with greater frequency in response to ongoing sea level rise and more intense storm events. The program – formally titled "Ruimte voor de rivier" – identified three general classes of strategies to expand flood plains in urbanized zones: (1) reestablish and widen the natural river channel; (2) create new spill zones for extreme flow volumes; and (3) elevate and reinforce hard infrastructure where development patterns cannot accommodate the natural river channel [26]. To carry out these strategies, several changes in the built environment of established cities – some more than a half millennium old – would be required, and these adaptations constitute the first principles of urban flood management in a climate changed world.

Drawing from both the technical strategies presented in the Dutch national water plan and adaptations that have unfolded in direct response to the implementation of these strategies, four core elements of urban flood management are exportable and, arguably, fundamental to adapting to a rising risk of flooding in all cities worldwide. I consider each in turn.

Widening the Floodplain. Engineered flood control infrastructure – such as levees, sea walls, and pumping systems – have the unintended effect of amplifying flood risk when stressed by water volumes in excess of their design capacity. When deployed in urban areas, such "hard" infrastructure stimulates land development in floodplains, increasing the size of the population exposed to the flood hazard should the infrastructure fail. Levees and flood walls further compound flood risk by elevating impounded waters, sometimes to heights greater than the structures to be protected, augmenting the hydraulic power of flood waters when these systems fail and reducing the time available for evacuation. Over time, these systems also tend to erode the capacity of the natural ecosystem to lessen flood volumes and intensity, which results from the degradation of wetlands in the floodplain that are starved of nutrients and sediment from the impounded river or tides.

The Dutch would confront this infrastructure-enhanced risk in response to a number of extreme storm episodes in the 1990s, supporting a growing view that ever-higher flood control infrastructure could not reasonably manage the risk of urban flooding in the face of rising oceans and intensifying storm events.

The Room for the River plan would not abandon the extensive use of levees alongside river systems but shift these structures further back from the river channel, widening the zones available for flooding. In some instances, river channels were deepened, increasing the capacity of the widened flood plain to transport greater volumes of water. The decision to relocate levee systems represents more than a new approach to flood management; it represents a sharp reversal from 800 years of human dominance over environmental systems in the Netherlands – a handoff in the responsibility for the flow of a river back to the river itself.

Relocation of Structures in Path of Widened Water Channels. The unavoidable consequence of widening a water channel to accommodate greater flow volumes is a need to devise new strategies for safeguarding urban structures now positioned outside of levees and other flood control infrastructure. For most situations, the only viable option is the relocation of homes, businesses, and other buildings beyond the widened floodplain. Of all the technical challenges presented by a changing climate to the field of urban planning, *retreat* – a small word with vast implications – is perhaps the most daunting because it is not a technical problem at its core but one of social equity. A legally established floodplain is a social compact obligating governmental intervention, often in the form of evacuation, temporary shelter, and assistance in rebuilding, when flooding events spill beyond the bounds of the floodplain. Given that such events are rare in a stable climate, with the very floodplain itself defined by historical storm and flow volumes, rebuilding in the aftermath of a rare event constitutes an acceptable collective expense. With the shift to an unstable climate – a circumstance contemporary disaster response programs and policies have until very recently failed to contemplate – the scope of this social compact must also shift to distribute the burden of retreat when flood defenses are removed.

For the Dutch, the relocation of families and businesses beyond expanded floodplains relies on established policies of eminent domain, through which property owners are legally compelled to relocate and compensated at what is determined to be a fair market rate. In some instances, property owners have requested that the government provide new land, as partial compensation, and new programs have been developed for doing so [27]. Such innovations in the established protocols of property condemnation constitute one of many new strategies that will be required to facilitate the relocation of populations on a much larger scale and, in some instances, entire communities. This emerging field of climate change adaptation will be explored further in Chapter 4.

Creation of Floodable Spaces to Lessen Pressure on Flood Defenses. During intense storm events, the city itself becomes a major driver of elevated water levels that threaten to breach flood protections. The expanse of impermeable paving and roofing materials designed to keep building interiors free of rainwater and streets operational during storms has the unintended effect of diminishing infiltration of rainwater into soils, an otherwise significant pathway for surface drainage. The common result is a shift from more than 80 percent rainwater retention in a forested area to less than 20 percent retention in a high-density urban environment. Highly impermeable cities, in this sense, generate their own floodwaters. The creation of floodable spaces within the fabric of urban neighborhoods reduces the total volume of water flowing over urban surfaces to rising rivers and other water bodies and increases the time required for waters impounded in these spaces to migrate to receiving waters, lessening the potential for flash flooding. This strategy has the associated benefit of expanding the availability of public greenspace in urban areas, often serving as new hubs for pedestrian and cycling networks.

The idea of a spillway to lessen pressure on levees and other hard infrastructure is not a recent innovation. The Dutch started constructing sections of levees that could be removed during extreme flooding events in the eighteenth century, diverting floodwaters into agricultural lands upstream from more densely populated zones [28]. What is new is the concept of carving out floodable spaces within the cities themselves, with some of the first examples of this strategy found outside of the Netherlands. In my own city of Atlanta, Georgia, for example, recurring flood events in neighborhoods undergoing rapid infill development have been mitigated through the construction of new greenspaces designed to capture and retain stormwater runoff during intense rain events. In the Historic Fourth Ward Park, constructed in 2008, a 5-acre section featuring a pond, walking paths, and a granite amphitheater can capture and store up to 9 million gallons of rainwater during storm events – safeguarding the surrounding neighborhoods from the flood risk of a 500-year rain event (Figure 2.5). While few concertgoers may realize that their neighborhood amphitheater could be fully submerged in the days following a performance, this is precisely the insight of the Dutch: In cities, adaptation must be cloaked as amenity.

Amphibiation. A final Dutch innovation inspired by the Room for the River plan is the idea that living with water sometimes means living on water. While houseboats have lined the canals of Amsterdam for more than a century, the planned construction of a neighborhood of floating structures incapable of operating under their own power is a very recent innovation. Completed in

Figure 2.5 Historic Fourth Ward Park, Atlanta, Georgia. Everyday Artistry
Photography/Alamy Stock Photo.

only the last two decades, several hundred floating and amphibious (capable of
floating during flood events) houses are now permanently moored atop or
alongside rivers and lakes in the Netherlands, principally in the largest cities of
Amsterdam and Rotterdam (Figure 2.6). The core innovation of floating communities extends well beyond a capacity to rise and fall with storm surges:
Amphibiation is a strategy for the achievement of a broader social and economic
resilience at the neighborhood scale [25]. As some percentage of the land
reclaimed from the sea in the Netherlands is returned to the sea, amphibiation
is offering a path to continued urban expansion and an emerging model for
ecologically autonomous settlement. In these experimental communities, climate
risk is the stalking horse for a shift toward more sustainable modes of living.

A compelling example of this idea can be found in Schoonschip, an
Amsterdam neighborhood composed of forty-six homes floating on the River
IJ in the central core of the city. Built atop floating platforms given buoyancy
and ballast from hollow concrete pontoons, each structure or "ark" is permanently moored to a series of jetties that create a set of alleyways stitching
together the neighborhood. While capable of rising with an influx of water
during storms, this climate-resilient aspect of Schoonschip is largely secondary
to other desired attributes, including affordability, the closed-looping of energy
and waste systems, and community engagement. Fashioned from common

Figure 2.6 Community of floating houses in Amsterdam, Netherlands. Wiskerke/
Alamy Stock Photo.

building materials and often designed to accommodate two families, the
multistory floating homes must be towed through narrow locks and canals
following construction, keeping their size modest and cost relatively low.
Requiring no land for occupancy, property taxes are assessed based on the
value of the structure alone, as well as a monthly mooring fee. The community
further enables affordability through the sharing of resources, such as collectively owned electric vehicles and e-bikes, and a floating community garden [29].

    With ample access to sunlight on the river surface, Schoonschip is powered
by an array of more than 500 solar photovoltaic and 60 solar thermal (for hot
water) panels installed atop most of the houses. The neighborhood is structured
as its own microgrid, allowing the generated power to be shared between
homes and any surplus to be stored in batteries underneath the structures.
Heat and air conditioning are generated by thirty heat pumps, which tap into
the river water itself as an energy source and, like the wealth of solar access,
leverage the inherent advantages of an aquatic setting. During the period of
a full year, Schoonschip generates more power than it requires, enabling the
neighborhood to sell surplus energy to the larger electrical grid and generate
a profit. As a fully autonomous microgrid, disruptions in power to the larger

region do not impact the neighborhood, providing a key dimension of resilience from extreme events that do not disable the community's own solar panels.

Like energy, drinking water is similarly collected and processed by the small community itself. Despite the availability of a municipal water connection, many of the homes are equipped with their own rainwater collection and purification system. Wastewater is separated within each home into two streams (greywater and sewage) and processed by a neighborhood-scale, floating organic digester. Biogases released from the wastewater treatment processing are collected and can be used as a supplemental power source by the community, creating an additional resource loop between water processing and energy production. Greywater collected from sinks and showers is presently returned to the municipal collection system but may in the future also be fully processed on-site [30].

For many residents, a final community benefit derived from the closely tethered design of Schoonschip is the most valued: regular social engagement with neighbors. Facilitated by the share economy of assets held in common, such as electric vehicles and a community garden, residents further plan weekly meals in each other's homes, arrange for regular "backyard" swimming time in warm weather, and use a dedicated app to request and share everyday items, often delivered in person within minutes [29]. Beyond this more structured interaction is the physical closeness of homes conjoined by a narrow jetty, with well-used porches and green roof patios – each family subject to the same ebbing rhythm of the river.

For some, the Dutch approach to amphibiation may seem to take the idea of flood resilience to an impractical extreme. While a large fraction of the global population lives in coastal regions, the development of floating neighborhoods within the active tidal waters of the world's oceans is not a viable strategy for extensive settlement. But a less extreme adaptation is entirely feasible from a marriage of the Dutch and American versions of the idea – one largely oriented toward permanent and the other toward temporary floatation – and would greatly expand the landscapes across which this strategy could enhance community resilience.

When moored atop inland water bodies outside of the main channel of rivers or active tidal zones, including river inlets, lakes, harbors, and canals, permanently floating structures are both feasible and affordable, particularly if linked into resource-integrated communities within the fabric of cities, allowing residents to leverage urban transportation, sanitation, and energy services to the extent required. A much larger region of land for the development of affordable and climate-resilient housing is available in the form of the expanding floodplain of rivers and inland water bodies. Neither the retrofitting of existing homes of suitable size for emergency buoyancy nor the construction of new homes capable of floating during extreme events is a technically infeasible or cost prohibitive adaptive response to flood risk with climate change. And these strategies carry the potential to work against a rising tide of human migration from land newly subject to periodic inundation.

For multistory buildings, both residential and commercial, an additional variation on the theme of amphibiation is the retrofitting of the ground floor to withstand flooding. Much like a levee overstressed by water levels exceeding its design capacity, many buildings require demolition after flood events due to a fracturing of structural supports as a result of water pressure. In the aftermath of Hurricane Sandy in New York, which impacted more than 80,000 buildings across a 51-square-mile flood zone in 2012, new building code standards have been adopted to require "wet floodproofing," which entails the installation of vents in exterior walls to permit water to flow into the ground floor during flooding, serving to equalize hydrostatic pressure on both sides of a wall and prevent structural damage [31]. Other common wet floodproofing strategies include the relocation of mechanical systems, such as heating and ventilation equipment, to higher floors or building rooftops, and a prohibition on ground floor or basement apartments in newly constructed residential buildings.

In practice, the Dutch strategy of making room for the river is more focused on the repurposing of land in the expanding floodplain than on the loss of critical urban space. With an eye toward adaptation as amenity, the Netherlands in the last two decades has reimagined and constructed massive floodplain expansion projects along three of Europe's largest rivers while simultaneously increasing public greenspace, pedestrian and cycling networks, critical wildlife habitat, and public access to waterfronts long barricaded behind imposing levees. These projects have required the displacement of homes and businesses in the path of unmanageable floods, but they have simultaneously created new public amenities for the urban population as a whole. New growth areas emerging in the form of floating neighborhoods represent an additional strategy characterized more by an enhancement than a diminishment in the quality of life for residents seeking a more sustainable mode of living. Novel adaptations of this idea, including the conversion of cargo containers into floating homes in post-Sandy New York, highlight the potential for urban redesign in response to climate change to achieve other significant policy goals, such as an expanded stock of affordable housing.

More radical in concept than engineering is the potential for amphibious housing and other flood management strategies to enhance the resilience of communities most at risk to flooding and least well prepared to adapt. Long displaced from higher ground to the most flood-prone bottomlands of cities, lower-income communities and, in most cities of the developed world, communities of color confront a much greater risk of flood inundation, adverse health effects, and economic losses than other urban populations. The scale of physical change required in cities to manage the suite of intensifying climate effects is daunting but brings with it an opportunity to work against historic disparities deeply imprinted on the urban landscape itself. If rapid ecological change and persistent racial and ethnic injustice constitute the most pressing societal challenges of the present century, as events of the first two decades render plain, our approaches to one must take direction from the other. Now is the moment to ally these aims.

The New Red Line

The sound of the collapsing condominium tower was closer to that of a building under construction than one undergoing a cascading set of structural failures. Described by surviving residents as intermittent hammering sounds, the first indications of what was transpiring at 1:00 a.m. on a warm summer night in Miami Beach were assumed to be the product of an inconsiderate neighbor's renovation project. Over a fifteen-minute period, the hammering sounds would continue, as overstressed steel rebar in the twelve-story building's concrete walls broke apart after forty years of bearing a weight in excess of its design capacity to support. Lacking the reinforcement of steel, the load-bearing walls in the subterranean parking garage would soon after fail, bringing down two-thirds of the building in seconds and entombing almost 100 residents in the dust-clouded rubble. As is often the nature of failure, the Surfside Champlain Tower collapse had unfolded at first gradually and then suddenly. For some, it was a first reckoning with the unforeseen hazards of living abreast a rapidly rising ocean.

While the official cause of the building collapse was structural failure associated with an undersized support system in violation of local building codes, there is little in the official investigation that sheds light on the timing of this failure. Beyond the simple passage of time, a growing physical stress on this building and others along Miami Beach is the frequency of soil inundation from rising seas. Situated only a few meters from the high tide line, a geotechnical analysis in the aftermath of the collapse would find the soils abutting the building's foundation to have experienced saltwater inundation at least three times as frequently in the years leading up to its collapse, in 2021, as at the time of the building's construction, likely contributing to structural damage of these support walls through corrosion [32]. Ranking among the deadliest structural failures in US history that were not a product of violent weather or terrorism, the Surfside Champlain Tower collapse may further mark the first large building brought down by the ebbing of a gradually but persistently rising tide.

A few city blocks inland from the site of the Surfside tower collapse, in the neighborhood of Little Haiti, the threat of sea level rise is assuming a different form. Too far from the beach to support much of a view, Little Haiti is so named due to the evolving mix of immigrants who have settled within the boundaries of what was once referred to as Lemon City. Subsequently incorporated into Miami and renamed to reflect the mix of Haitian, Bahamian, and other Caribbean immigrants who account for the majority of the neighborhood's population, Little Haiti remains an important beachhead for Afro-Caribbean culture within the larger fabric of Miami, as attested to by the density of restaurants, music venues, art galleries, and festivals reflective of these nationalities. Little impacted by the construction booms of past decades, the neighborhood in recent years has attracted new investment and redevelopment due to the very attribute that has insulated it from gentrification throughout the span of its history: At 7–14 feet above sea level, Little Haiti commands some of the highest ground in Southern Florida.

That a beachside building collapse could further accelerate interest in Little Haiti and other predominantly Black and brown communities of Miami is very much on the mind of community leaders in these neighborhoods. For, regardless of the degree to which sea level rise destabilized the Surfside Champlain Tower, the growing perception of risk along the coastline appears to be, at long last, influencing property values in Miami. Only in recent years have real estate trends reflected a gradual diminishment in the appreciation of residential property in low-lying areas of Miami compared to the value of property further from the coastline and situated in neighborhoods at higher elevations. The result in long-neglected neighborhoods like Little Haiti has been a spiking of land values, with average home resale values rising by more than 200 percent between 2012 and 2021, compared to about a 40 percent increase in home values for Miami Beach during the same period [33]. Coupled with rapidly rising property values is a disproportionate rise in the rate of rent hikes and evictions in higher elevation neighborhoods, with rents in some apartment buildings doubling overnight following the sale to a new owner [34]. As observed by a lawyer representing displaced clients, "We're seeing developments in areas we've never seen before. If anyone tells me developers bought there for any reasons other than climate gentrification, I'd call them liars" [35].

The accelerating displacement of lower-income and, predominantly, Black and Latino families in Little Haiti is an ironic coda to the historical forces that gave rise to an ethnically and racially diverse enclave several miles inland from the tideline. As Miami developed in the early twentieth century, the high ground of Lemon City was more affordable than property closer to the water, enabling those priced out of other parts of the city to find housing further inland. By the 1930s and 1940s, the economic impediments to purchasing property in lower-lying, more desirable neighborhoods of Miami would be further constrained by access to mortgage lending. Characterized in popular parlance as "redlining," differential access to credit by race was an inherent feature of the modern mortgage structure when first conceived during the New Deal era in the United States. And it is the effective ghettoization of redlined neighborhoods by lending institutions that almost a century hence most sharply delineates the physical contours of climate vulnerability.

In need of a macroeconomic stimulant during the Great Depression of the 1930s, the US Congress approved the creation of new national banking entities to make home ownership affordable to more Americans. Doing so would require that the lending period be increased from a standard of less than ten years to more than twenty – the effective creation of the popular thirty-year mortgage – and that the federal government compensate local banks if borrowers failed to repay their loans, serving to lower eligibility requirements for such loans and greatly widening the pool of potential borrowers. While a succession of new statutes would be required to achieve these objectives, one of the resulting entities – the Home Owners' Loan Corporation (HOLC) – would institute a practice of property appraisal that would systematically exclude non-white families from home ownership for decades to follow and, in important respects, to the present moment.

Charged with the responsibility of refinancing mortgages for borrowers in default on loans issued by private banks, the federal HOLC needed a standardized approach for assessing the creditworthiness of borrowers in all regions of the country. The technique devised to achieve this end was a rating system through which cities were subdivided into zones and each zone was assigned a grade corresponding to the assumed likelihood of property value appreciation over time [36]. Zones characterized by newer construction, available land for additional housing, and ethnic and racial homogeneity – factors deemed most predictive of appreciating property values over time – were assigned a grade of "A" and shaded green on an accompanying set of maps. Zones characterized by older construction, a mix of residential and nonresidential land uses, and a significant presence of non-white residents were most often assigned the lowest grade of "D" and shaded red in maps used to assess lending risk. No data on historic property value appreciation or the rate of loan defaults by zone was considered in assigning these grades.

Soon adopted by private banks in addition to federal lending institutions, the HOLC framework for assessing creditworthiness would result in a systematic denial of loans to homeowners in the red or "redlined" zones – giving life to among the most effective economic policies yet devised for empowering the upward economic mobility of one group of borrowers at the direct expense of another. Despite the decommissioning of the HOLC by the early 1950s, its framework for assessing neighborhood creditworthiness was already widely in use by private lenders and would not be deemed unlawful for another three decades. As I write, more than a handful of private banks remain in active litigation over the issue, but the extent to which redlining is occurring today is almost beside the point: Many decades of unequal access to private lending has excluded non-white families from the most basic tools afforded by society for wealth accumulation and intergenerational transfer. Almost a century later, neighborhoods redlined in the 1930s remain poorer, less structurally sound, and served by lower-achieving schools than adjacent zones assigned a higher grade. And they remain far more vulnerable to the extremities of climate change [37,38].

To overlay a HOLC creditworthiness map from the 1930s on present-day neighborhood boundaries in almost any large US city is to, more often than not, delineate the areas of highest flood risk and heat exposure. Across thirty-eight of the largest cities, the share of homes at risk of flooding in neighborhoods assigned the lowest grade by the HOLC is more than 20 percent greater, on average, than the share of homes at risk of flooding in neighborhoods assigned the highest grade [39]. A survey of more than 100 US cities finds substantial differences in cooling tree canopy by HOLC zone, with redlined neighborhoods exhibiting less than half of the canopy cover as neighborhoods characterized as the most creditworthy almost a century ago, resulting in markedly higher levels of heat risk [40]. The long-ago invented hazard of race in these zones is today the now effectuated hazard of flooding and heat exposure due not to any differential in the pace of climate change between these zones but to the creation of the zones themselves. To be born into a Black family in the United States is to confront a much greater probability of a life in poverty, less access to nutrition and adequate healthcare, and poorer baseline health [41,42] – all of which elevate the risk of injury or death from the physiological stresses of climate change.

In Little Haiti, the zone designated by the HOLC as Class D or "Hazardous" in the 1930s fully encompasses the contemporary boundaries of the present-day neighborhood (Figure 2.7). In contrast to the vast majority of redlined neighborhoods across the United States, which are more likely to occupy the most flood-prone land in cities, Little Haiti sits atop what is quickly becoming some

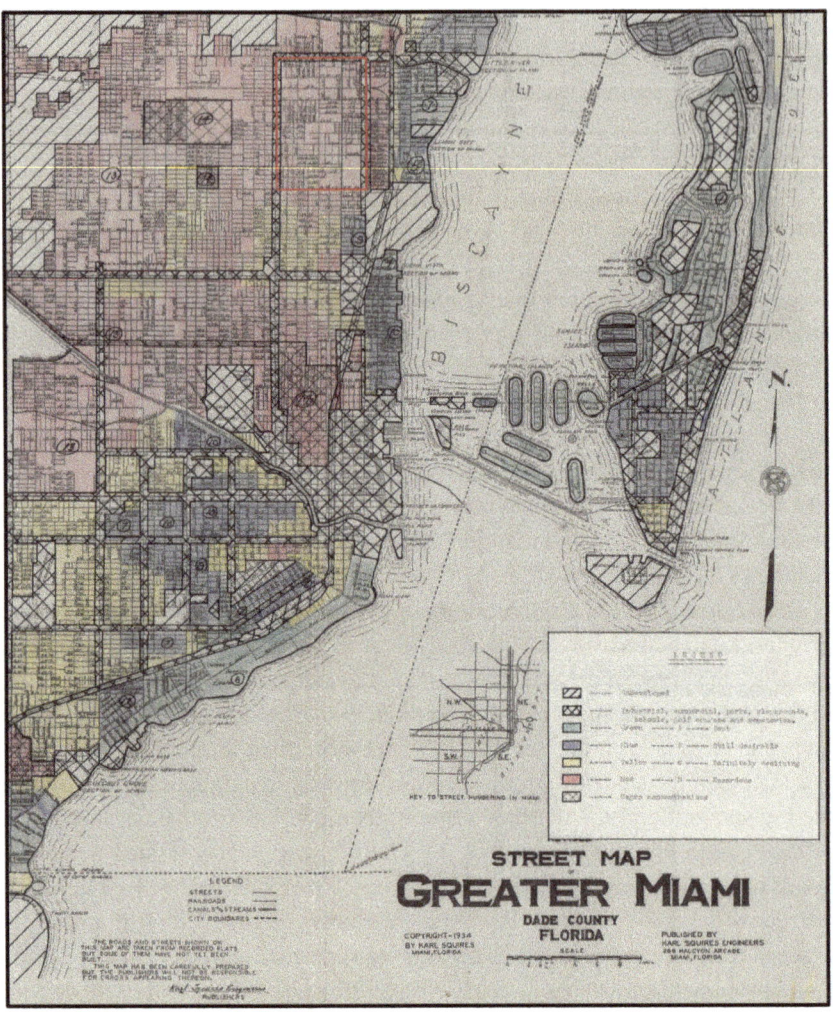

Figure 2.7 Approximate location of the Little Haiti neighborhood (outlined in red) in the 1934 Home Owners' Loan Corporation (HOLC) map for Miami, Florida. US National Archives, https://catalog.archives.gov/id/85713715.

of the most desired property in Miami for redevelopment. But its residents may be no more likely to avoid harm or displacement due to climate change than other residents of redlined neighborhoods, given that rapidly rising land values and rents are forcing those least well positioned to manage the rising risks of climate change out of the most flood-resilient zone. Like all cities, Miami will need to invest heavily in climate adaptation to lessen the vulnerability of its residents to intensifying impacts in the form of extreme heat, flooding, and drought. The experience of Little Haiti highlights a core principle of urban climate adaptation: Communities with the greatest vulnerability to climate change, often as a result of discriminatory governmental policies, must be prioritized in time and in the scale of governmental action for resilience-enhancing investments, including policies protecting against displacement. I refer to this principle as "least-first," and it is fundamental to the idea of radical adaptation.

Those communities least well protected from climate extremes by ecological bulwarks, such as high ground and extensive vegetative cover, must be prioritized for climate adaptation investments designed to lessen flood risk, moderate heat exposures, or otherwise enhance neighborhood-wide resilience to climate-related exposures rendered more intense by the state of the local ecosystem. Those communities least physiologically adapted to climate extremes, as a product of age, compromised health, or recent migration from different climate zones, must be prioritized for governmental interventions designed to lessen climate-related exposures and improve baseline population health. And those communities least equipped with adaptive resources to manage climate-related exposures, such as housing raised above base-flood elevations, mechanical air conditioning systems, or access to real-time information on climate hazard risk, must be prioritized for interventions that enhance individual or community-scale adaptive capacity.

While this principle of least-first may seem at first blush uncontroversial, it is strongly at odds with the earliest actions of many large cities seeking to adapt to emerging climate risks. Consider, for example, the highly ambitious campaign of New York City to add one million trees to the city's canopy in less than a decade – a campaign undertaken, in large part, to lessen heat exposure and flood risk across the city. With about 80 percent of these trees targeted to parks and other public land across New York, it has been the greenest neighborhoods of the city that have received the most new tree canopy, to the exclusion of lower-income neighborhoods that tend to have less public greenspace [43]. Likewise, one of the largest climate adaptation awards yet granted by a US federal agency – more than $140 million for flood resilience planning in New Orleans in the aftermath of Hurricane Katrina – would be targeted in its entirety to the Gentilly District, encompassing neighborhoods with an average household income above the median for the city as a whole.

The least-first principle is radical in concept in that it inherently acknowledges that cities are not positioned to act everywhere equally – that public investments must be prioritized. Beyond its imperative that the least affluent communities be prioritized in the expenditure of public funds for climate adaptation, the least-first principle further militates against the idea of an equitable distribution of funds, wherein each planning district in a city receives investment equal in proportion to its population size. While not in conflict with the idea that cities must be redesigned across their full extent to manage climate risk, the least-first principle requires that these investments be strategically deployed to the most vulnerable communities first. For municipal governments confronting the challenge of rapidly rising climate exposures, the first question to be answered is not *how* to intervene but *where*, as the socio-spatial context must inform the strategic approach. The least-first principle is in this sense a precondition to adaptation, and it requires municipal governments to be *reparative* in their approach to climate adaptation.

An imperative to prioritize the needs of the most vulnerable and historically marginalized communities in climate adaptation planning – often referred to as *climate justice* – increasingly is being codified into climate management programs of governments and nongovernmental organizations. While many plans to date invoke the notion of climate justice, a more limited but growing number hardwire these goals into the direct disbursement of adaptation investment and other resources. In January 2021, the US federal government committed under the Biden administration to direct a minimum of 40 percent of all federal spending on climate adaptation and clean energy projects to disadvantaged communities – a sizable fraction of national investments in this area [44]. Specific criteria have been issued on which communities qualify for augmented expenditures and specific funding initiatives, such as the Flood Mitigation Assistance Program, and include historical and projected losses in buildings, population, or agricultural productivity across all US census tracts in response to climate-related events. Titled the "Justice40 Initiative," all relevant agencies of the US government are required to demonstrate annual compliance with this requirement in budgets and policy programming.

With a more expansive set of policy tools to redesign urban landscapes, municipal climate adaptation programs are most well positioned to direct climate resilience efforts to the most vulnerable neighborhoods in cities. To date, however, only a small number of US cities has incorporated least-first principles into climate adaptation plans in a meaningful way. A survey of 100 US cities found that only a little more than half had developed climate adaptation plans of any sort. Of these, only twelve cities had included measurable policy actions to prioritize adaptation investment in the most vulnerable communities [45]. One of these cities, Providence, Rhode Island, has developed what may be the nation's first stand-alone climate justice plan, through which the city's most climate-vulnerable communities – referred to as "frontline communities" – are identified and prioritized for investments in green infrastructure, microgrids powered with renewable energy, home weatherization, and regular monitoring of neighborhood-scale environmental quality [46]. Climate adaptation strategies piloted in these frontline communities may then inform a more spatially comprehensive set of actions over time. More cities should follow this impressive lead.

As a product of the old red line, delineating zones permitted for settlement by disfavored racial and ethnic groups and found in some form across all cities, a new red line is emerging: The communities least well equipped to manage flooding and heat exposures exceeding our physiologic limits will be the first to endure these exposures in large cities. The combination of social injustice and climate risk is particularly acute in cities as the physical characteristics of the city itself amplify the extremity of heat and the intensity of flooding in direct response to adaptive capacity. The existence of this red line is instructive in terms of both the strategies that are required to combat extreme heat exposure and flood risk and how these strategies must be staged. Our efforts must be more than ecologically restorative; they must be radical – in both degree and deployment.

Radical Adaptation to Flooding

My aim in this chapter has been to highlight several dimensions of the threat to cities posed by rapidly changing patterns of rainfall and global sea levels. The first of these is that no city is inoculated from an increasing risk of extreme flooding events. While coastal cities confront an enhanced threat of flooding from accelerating sea level rise, inland cities confront the risk of flooding from more intense rain events that are amplified by the expanding built environment of the city itself. In concert with the rising threat of extreme heat events, every major city worldwide will need to develop new and expanded approaches to flood management, and no city to date is fully prepared for the intensity of extreme rainfall events that are occurring with greater frequency.

To be effective, these new approaches to flood management will need to be radical in at least two respects. First, much like the adaptive response to heat risk, the resurfacing of cities required to manage the growing risk of flooding must be both granular and spatially *dispersive*. No neighborhood or land parcel can be exempted from the need to enhance on-site rainwater collection and infiltration while substantially diminishing stormwater runoff. In contrast to large and often centralized public works projects, such as municipal wastewater treatment facilities, stormwater collection and management in a climate changed world cannot be accommodated through traditional hard infrastructure alone, and it cannot be accommodated without addressing the impervious nature of urban construction neighborhood by neighborhood. Most every impervious area of the city over time must be resurfaced or reduced in size to respond to the magnitude of risk presented by a growing intensity of storm events and the expanding volume of adjacent water bodies.

Water-responsive climate adaptation further must be radical in its deployment. Often characterized by the highest levels of exposure to extreme heat and flood risk, combined with the lowest adaptive capacity to manage these exposures, historically marginalized communities, by race, ethnicity, and/or income, are appropriately described as frontline communities for climate change. Publicly funded adaptation efforts must depart from the established paradigm of directing climate-adaptive infrastructure to the areas with the greatest political payoff at the expense of those with the greatest socioeconomic need. Moving forward, these strategies should aim to be *reparative* in their deployment. The least-first principle tells us where to act first and, in so doing, what it is that must be done. It is a simple yet powerful device for jump-starting climate adaptation planning in cities. No measure of resilience can be achieved without it.

3

Water (Too Little)

While not the first major city to see its water reservoirs depleted by a shift in climate, Cape Town, South Africa would be the first to identify in advance the precise day the taps would run dry – and the first to refer to this date as "Day Zero." The origin of this phrase is not entirely clear. In the military, Day Zero has long referred to the date of induction into training, most often for those conscripted into service. More recently, the phrase has been coopted by computer hackers to refer to the date upon which a newly exploitable vulnerability is identified in a string of code. If there is a common thread to these variations on the idea, it is this: Day Zero is the day one set of established and predictable conditions ceases to operate and another, far less desirable and predictable set of operating conditions takes hold. For the unwilling conscript, user of software, or resident of a large city, Day Zero is something to be avoided.

The first major city to reach Day Zero for a large fraction of its population was São Paulo, in 2015. As recounted in the Prologue to this book, several years of persistent drought conditions – fueled in part by the accelerating loss of rainforest across the western and central reaches of Brazil – almost fully depleted an extensive system of reservoirs in one of the wettest zones of the planet, forcing municipal water delivery shutoffs for an undisclosed number of residents over weeks at a time. While municipal drinking water would continue to flow for some fraction of the population, no prior event had resulted in such a prolonged disruption of water delivery for millions of residents from regional reservoir depletion.

The first and only major global city at the time of writing to have fully exhausted regional drinking water reservoirs and, in so doing, see Day Zero come to pass is Chennai in 2019. As briefly described in Chapter 2, this southern Indian city with a population exceeding 10 million was unable to deliver drinking water to residents following several years of delayed and deeply diminished monsoon seasons. To sustain the population, water was

Figure 3.1 Delivery of water via tanker train in Chennai, July 2019. P. Ravikumar/ Reuters.

delivered daily via long tanker trains and strictly rationed to snaking, over-heated, and sometimes violent queues of residents with plastic buckets – an emergency operation that would continue for four months [1] (Figure 3.1). Chennai is only one of several Indian megacities confronting chronic ground-water depletion, with the rising demand for water across the country projected to *double* by 2030. An estimated 40 percent of the population – more than half a billion humans – may lose regular access to drinking water in less than a decade [2].

   But it is arguably Cape Town, more than any other city confronting a severe water crisis, that most clearly foretells the magnitude of risk confronting the most populous cities of the world, regardless of location in Global North or South. In contrast to São Paulo or Chennai, Cape Town has not intensified the effects of long-term drought through an overexploitation of groundwater resources, the loss of regional rainforest, or the extensive destruction of wetlands. While both Cape Town and Chennai experienced an approximate 50 percent reduction in rainfall leading up to their respective Day Zeros, Chennai's capacity to manage drought was and is greatly compromised by the loss of more than 80 percent of its natural wetlands in recent decades [3]. Combined with a long-term trend toward falling water tables, this substantial loss of natural water storage capacity has accelerated the depletion of regional reservoirs there and in other cities experiencing unchecked population growth.

Cape Town would also institute serious water conservation measures in response to diminished rainfall and falling reservoir levels much earlier than did São Paulo or Chennai. In the aftermath of a more limited drought in the early 2000s, Cape Town invested in expanded reservoir capacity, increasing the total drinking water supply by 17 percent. In 2015, the first year of the drought, Cape Town received international recognition for its water conservation efforts from C40 Cities, a global organization focused on urban climate management. Out of fifteen cities receiving such awards at the United Nations Conference of the Parties (COP 21) meeting in Paris that year, Cape Town would be the only city worldwide recognized for its water conservation efforts, which entailed free household plumbing repairs to address water leakage issues and an aggressive block pricing scheme for water consumption during periods of drought [4]. At the start of the drought, the highest consuming household would pay four times as much per liter of municipal water as the lowest consuming household; one year into the drought this differential would increase by a factor of 17 [5].

Years in advance of the projected arrival of Day Zero, Cape Town had established itself as a global leader in the realm of water conservation. Indeed, Day Zero would first be recognized as a threat to Cape Town due not to any lack of preparation for its arrival but as a direct byproduct of these preparations: The very idea of a doomsday clock for water scarcity was created by water managers in a city that was well focused on such an eventuality and seeking a device to communicate this threat to the public. In stark contrast to government officials in São Paulo, who concealed their efforts to manage the water crisis, or those in Chennai, who adopted almost no public conservation efforts in advance of discontinuing public water service, Cape Town can be seen as an example of highly effective water stewardship and conservation management. And it is precisely this aspect of the South African water crisis that should be most unsettling to municipal governments elsewhere.

The recent experience of Cape Town, the first city to pinpoint Day Zero on the calendar – April 12, 2018 [6] – and to marshal every resource available to a nation with a capacity for planning and disaster management on par with cities of the developed world, is most revealing of ways in which drought is altering the trajectory of human settlement perhaps more substantially than any other mode of climate change. Water scarcity does not conform to coastlines, is not limited by latitude, and is far less responsive to the physical design of the city than extreme heat or flooding. More than any other dimension of climate change, too little water is the principal driver of the great climate migration already underway – a mass resettlement that is contributing to a destabilizing of political systems on every inhabited continent. To bet on the near-term impact of water scarcity on the lives of urban residents worldwide, in cities already experiencing drought and those that are not, is to wager with the advantage of the house.

City As Wellspring

Little about the label reveals the unique provenance of Singapore's NewBrew Tropical Blonde Ale. The colors are pastels that one associates with a tropical setting; each can design depicts a different water-related activity, such as kayaking or running along a beach; the reported alcohol content is 5.1 percent. Upon its release in April 2022, the beer received strong local reviews and sold briskly, with many stores selling out once word of the new approach to beer making received attention in regional and international media. One reviewer described the beer as "smooth, with a toasted honey-like aftertaste"; another, feeling inspired, recommended the beer as an ideal choice for anyone seeking to "get pissed" over the weekend [7]. To this end, the most prevalent theme of media reviews was not the afternotes of the beer but the novelty of its distillation: NewBrew's Tropical Blonde Ale is composed chiefly of recycled urine.

If the idea of reclaiming municipal sewage for the brewing of beer is off-putting, it is important to note that virtually all commercially available beer is derived from human wastewater, only with one additional time step than is the case for NewBrew. For any city that is located downstream of another, the most common source of drinking water is composed, at least to some extent, of an upstream city's treated municipal wastewater. Once drawn from surface water, this recycled municipal wastewater is again treated to drinking water standards and supplied to urban residents and industries, such as breweries. Singapore's NewBrew shortens this process by directly treating and reusing the city's own wastewater, rather than drawing on the treated wastewater of an upstream city. To enable this closed-loop approach, NewBrew uses a multiple-step filtration process, including reverse osmosis, to deliver pure water to the brewing process. In so doing, NewBrew is making use of a water input that is of a far higher quality than most any source of municipally treated drinking water worldwide. The reality may be hard to swallow, but beer derived from urine is likely the purest beer you will ever have.

Singapore would become one of the first countries to recycle municipal wastewater for drinking water in the early 2000s, following a period of drought that severely threatened drinking water supplies. Roughly the size of Chicago but with twice the population, Singapore's land area is simply too small to amass from rainfall the quantities of drinking water required by its domestic and industrial demand. Having ruled out desalinization of surrounding sea-water as too expensive and energy-intensive, the national water supply authority has sought strategies to supplement freshwater supplies delivered under treaty from nearby Malaysia. Today, 40 percent of the country's drinking water is derived from wastewater recycling, shoring up the national water supply during periods of drought [8].

Singapore is one of numerous global cities situated in wet regions of the planet but still facing the possibility of a future Day Zero. Jakarta, London, Miami, and Tokyo – all cities characterized by tropical to temperate climates – are also increasingly at risk of exceeding the local demand for freshwater supplies in a climate changed world [9]. While the local stressor on water availability differs by region, ranging from a loss of groundwater to the salinization of aquifers from rising seas, the one constant variable is a shift over time toward greater extremity in the pattern of rainfall. For many cities, this shift is resulting seasonally in more water than present infrastructure can capture and store in wet periods and less water than is needed for industry, agriculture, and drinking water in dry periods. As this shift in rainfall patterns carries more of a global signature than a local one, municipal, provincial, and national governments generally lack the policy or technological capacity to much influence these patterns.

The major variable at play is a seemingly small but consequential shift in planetary water storage from terrestrial to atmospheric reservoirs. As introduced in Chapter 2, the Clausius–Clapeyron rate quantifies how much additional surface moisture can be evaporated and stored by the atmosphere with each 1°C (1.8°F) rise in temperature. Its core implication is that a warming world, as more moisture is shifted from the land surface and the oceans to the atmosphere, is increasingly a drier world, at least on the planetary surface. As we approach 1.5°C (2.7°F) of global warming, we must acknowledge that a reduction in global surface water between 7 percent and 14 percent is not the manifestation of a drought – a short-to-medium-term shift in precipitation patterns – but a lasting change in water availability.

This shift is already upon us. Analysis of rainfall records since the late nineteenth century reveals an extensive and long-term progression toward drought events of greater frequency and duration across every populated land mass on Earth [10]. The physical area of the planet experiencing drought conditions in any year today, in response to 1°C of warming, is estimated to be about 20 percent greater than observed just a few decades ago (1970s–1990s) and is projected to reach a land area 50 percent greater than historical extents in response to 1.5°C of warming [11]. These numbers indicate that a half degree of additional warming, the path we have committed to over the last decade alone [12], is estimated to increase the area of the global land surface experiencing drought conditions in any year by a factor of 1.5. An additional half degree beyond that – reaching 2°C (3.6°F) of global warming – would increase global drought zones by another 7 percent, while increasing the annual frequency of drought episodes by almost 70 percent relative to the recent past [11].

Given the vast disparity in water availability between climate futures separated by only a degree in mean global temperatures (1°C–2°C), how likely are we to cross the 2°C threshold this century? The answer to this simple question – arguably the most consequential of our lifetimes – is that there is no demonstrated pathway to avoid a 2°C rise in global temperatures by the end of the century [13]. Projections by the United Nations Environment Programme presently foresee a 2.7°C (4.9°F) rise in temperatures by 2100 [14], accounting for all realized and planned national reductions in emissions. Even the attainment of net-zero emissions by a large number of industrialized nations would result in a rise in mean global temperatures of 2.2°C (4°F) by the end of the century. In short, the weight of the most recent evidence, even accounting for emission-reduction commitments made under the Paris Agreement that are already off target, plainly reveals that we are headed for 2°C of warming this century, quite possibly by 2050, and that this warming is likely to limit global freshwater availability to levels unknown in modern human experience [14].

With no foreseeable path to keep mean global temperatures below 2°C this century, and thereby avoiding sharp reductions in global freshwater availability, cities in all regions of the planet must critically assess long-established practices for the retention and use of freshwater. To this end, the city must be redesigned and its natural bulwarks restored to play an active role in retaining rather than repelling the more limited quantities of precipitation received each year. In some regions, the institution of meaningful water conservation methods, during and between periods of limited precipitation, may extend regional water security well into the future; in others, the shift toward greater water scarcity will be too extreme to support urban populations and regional agriculture at historical levels. Effective adaptation to growing freshwater scarcity will entail the enhanced collection, intensive recycling, and sharply pared-back use of water.

City As Receptacle

The most basic approach to addressing increasing water scarcity is to retain more of the limited quantities of rainfall received. Of the two principal drivers of meteorological drought – a reduction in seasonal or annual precipitation and an increase in water losses through runoff and evaporation – the loss of received rainwater is the driver over which humans have the most control, at least in the near term. Extensive impervious materials with urbanization, a denuding of natural vegetation, and a lack of shade from tree canopy greatly enhance rates of water loss through runoff and evaporation, limiting the quantity of water absorbed and stored by soils within and in proximity to cities.

The vast expanse of cities occupied by surfaces impermeable to rainwater, with most of these surfaces dedicated to the operation and storage of vehicles, provides a singular statistic needed to understand the dilemma of drought adaptation in urban areas. Across the most populous US cities for which different land cover types have been consistently mapped – a list including Chicago, Denver, Houston, New York, and Los Angeles – the area of streets, parking lots, and buildings as a percentage of total land area exceeds 50 percent [15]. In other words, more than half of every square meter of these cities, on average, is impermeable to rainwater. By contrast, the total area occupied by grass and tree canopy – surface materials conducive to the infiltration and storage of rainwater – averaged 34 percent, just one-third of the urban land area. With so much more land dedicated to repelling rainwater rather than collecting it, many large cities are engineered to promote rather than ameliorate drought.

It is not surprising to observe that the most effective approach for retaining rainwater in cities – an extensive restoration of green cover at the neighborhood level – aligns directly with the most effective approaches to managing extreme heat and flooding. For the principal driver of climate-related impacts in cities to date is not a failure of our levees or our storm sewer systems or the lack of mechanical air conditioning in every home but rather the dismantling of the innate ecology entombed beneath its buildings and streets. As climate-related stressors in cities intensify with continued global-scale warming, a restoration of local ecological capacity for the regulation of stormwater and the moderation of heat will not bring about a restabilized climate, but it can lessen climate impacts as global emissions presumably peak and commence a decline this century. In this sense, ecosystem restoration in cities is an adaptation bridge – arguably the only one we have.

Combined with a need to lessen the quantity of rainwater repelled by the waterproofed surfaces of cities is a need to store it for irrigation and other non-potable uses. More than a third of all water use in US cities is dedicated to the watering of residential landscaping – a statistic that rises to 60 percent in some of the most arid cities of the American west [16]. As both surface and groundwater resources rapidly diminish in the western US, some of the most

Figure 3.2 Replacement of banned grass with artificial turf in Las Vegas neighborhood. Matt McClain/Washington Post/Getty Images.

restrictive water conservation measures to date are focused on outdoor watering. At the time of writing, Los Angeles – confronting a drought of historic extremes that signals the ongoing shift to a more arid climatic regime – has restricted all outdoor watering to twenty minutes a week [17], a prohibition that is rapidly browning lawns and withering residential orchards across the city's intensively irrigated neighborhoods. In response to similar drought conditions, Las Vegas has enacted a law requiring all "non-functional" lawn areas to be removed entirely (Figure 3.2) [18]. Perhaps for the first time since their founding, major cities of the American west are slowly reverting to the ecological constraints of their desert settings.

To make do with less, cities will need to become receptacles for the capture and retention of rainfall. Rainwater collection and storage, using the built environment itself for these purposes, is a practice as old as city making. The earliest Greek and Roman ruins, for example, reveal the practice of harnessing roofing systems as water collection devices and channeling intercepted water to reservoirs within and beneath buildings. The Greco-Roman *impluvium* was a shallow receptacle positioned at the center of a central atrium in private homes, through which rainwater was received directly from an inwardly pitched and open roof referred to as the *compluvium* (Figure 3.3). Extensive

Figure 3.3   Ancient Roman house with *impluvium* in Pompeii, Italy. Roger Viollet/ Getty Images.

excavations at Pompeii reveal a highly engineered system in larger homes, in which water collected by the impluvium was channeled through a sand filter to a subterranean cistern beneath the structure, allowing for the storage of filtered and cooled drinking water. Ruins of Mayan cities and other Mesoamerican groups likewise exhibit engineered systems for the collection and long-term storage of water in an arid climate, often making use of the same stone quarries excavated for temple building to serve as water storage reservoirs [19].

The redesigning of cities to capture and store rainwater, and to ultimately recycle it, is referred to as *sustainable urban drainage systems* or SUDS. The simple precepts of this approach to urban hydrology, that cities most efficiently manage water resources through mimicry of ecological systems, can be traced to Ian McHarg, a key figure in the environmental planning movement of the 1960s and the author of the seminal text on the topic, *Design with Nature* (1969). McHarg would propose at that time, and carry out through his design practice, the radical idea of replacing stormwater pipes underneath the road with vegetated drainage ditches alongside it. When designed as part of a larger network of drainage ditches – now referred to as bioswales – and stormwater collection areas – now referred to as floodable parks and urban wetlands – such simple design solutions can handle volumes of water equivalent to subterranean storm sewers at a far lesser cost while generating an array of co-benefits in the form of recharged groundwater, reduced surface water contamination, air temperature regulation, expanded recreational greenspace, and enhanced property values.

The first component of SUDS is rainwater harvesting. Never lost in cities of North Africa, the Middle East, and India, the ancient practice of rainwater collection and storage is making a strong resurgence in areas of the developed world confronting unprecedented drought conditions. While the expense of treating harvested rainwater to drinking standards at the scale of individual homes is generally prohibitive, drinking water supplies can be greatly extended by using rainwater for irrigation and other greywater purposes, such as the flushing of toilets. Incentivized by rising rates of municipal water service in arid cities, a growing number of homeowners are affixing gutter downspouts to rainwater reservoirs much larger than the traditional rain barrel, in some instances installing underground cisterns capable of storing hundreds to thousands of gallons of captured water. The potential for such systems to reduce the demand for municipally treated drinking water is impressive, with studies in both arid and wet regions finding 30 percent to almost 50 percent of residential water demand to be met with household rainwater harvesting [20,21].

Outside of North America, rainwater harvesting is a widely used strategy to reduce the demand for drinking water and lower household expenditures on rising water bills. Confronted with extreme water insecurity, many large cities in India mandate the installation of rainwater harvesting systems in residential buildings [22]. Australia has become a global leader in rainwater collection and use, with more than 30 percent of single-family homes having installed a system and with half of these systems connected to indoor non-potable uses, such as toilet flushing. In recent years, one-third of all new residential buildings in Germany – among the wettest countries in Europe – feature rainwater harvesting systems. France, Spain, and the UK all provide government subsidies for such systems, while Sweden is experimenting with rainwater harvesting from roadways. Thailand and China have provided millions of harvesting systems to residents at no cost; Japan and South Korea offer generous subsidies to support rainwater harvesting to conserve municipally treated drinking water [21,23].

For captured rainwater not stored or reused on site, the second component of SUDS is infiltration to groundwater. Raingardens, floodable parks, urban wetlands, and engineered infiltration basins are all devices designed to return rainfall to the local water table and in so doing lessen the volume of water diverted into the urban storm sewer system. As most urban sewer systems combine both sewage and stormwater runoff, stormwater volumes exceeding system capacity often result in the discharge of combined waste and rainwater directly to natural waterways with little or no treatment of contaminants – a startling legacy of the preindustrial city through which raw sewage is flooded into lakes, rivers, and neighborhood streams frequented by children, family pets, and local wildlife. The diversion of rainfall away from storm sewer drains via bioswale channels linked to floodable spaces offers an elegant solution to what remains a serious water pollution issue in most every major city worldwide.

The replenishing of local water tables through rainwater infiltration works against the dehydration of cities produced from an excess of impermeable surfaces in the form of streets, parking areas, and buildings. As explored in Chapter 2, the diversion of rainfall from local infiltration has the effect over time of compressing urban soils and measurably lowering the elevation of the city, further enhancing the potential for flooding. The retention and infiltration of rainwater where it falls enhances the capacity of the urban environment to cool itself through evapotranspiration and to support more extensive tree canopy without the need for regular irrigation. For cities dependent upon groundwater for drinking water supplies, the benefits of greater rainwater infiltration within the city itself are self-evident.

The diversity of strategies for enhancing the permeability of cities today ranges from the ancient to the experimental. With a matrix of sand and soil to function as a natural mortar, the brick-laid streets of nineteenth-century Chicago yielded a far greater porosity to rainfall than the ubiquitous asphalt paving of the city today. In recognition of the stormwater burden created by extensive impervious surfaces, Chicago, Detroit, and Washington DC are once again resurfacing miles of alleyways with reclaimed brick and a contemporary form of cobblestone: the permeable paver. Widely used across Europe and a growing number of Asian and North American cities, the most recent gener-ation of permeable pavers enables infiltration in the joints between intercon-necting blocks and through the permeable interior of the paver itself (Figure 3.4). When properly installed and maintained, permeable pavers can increase the infiltration of rainwater along roadways and parking lots by up to 100 percent, greatly lessening the volume of runoff conveyed by these streets to combined sewer systems and augmenting the local retention of water [24].

More effective than the replacement of impervious materials with porous paving is the growing practice of shrinking the overall footprint of impervious surfaces in cities. Perhaps the most long-lived version of this approach is the use of soil and plants atop buildings as a natural means of roofing insulation,

Figure 3.4 Permeable paving in Albuquerque, New Mexico. Ken Hawkins/Alamy Stock Photo.

a practice dating to the use of sod roofs in Viking settlements of the ninth century. In recent decades, the contemporary green roof, in which a wide array of usually low-stemmed and drought-resistant vegetation is grown in a soil media atop a flat and or moderately pitched roof, has been increasingly used to lessen building runoff, moderate roof-level temperatures, and redeploy roof area in dense cities as elevated garden space. By one estimate, more than $1 billion is invested each year in green roof development worldwide, with the number of installations growing by 17 percent annually [25].

A more direct connection to the urban water table than is achievable with green roofs is the repurposing of roadway paving for rainwater collection. In protest of the over-allocation of urban space to vehicles in North American cities, a group of activist planners in San Francisco converted a single on-street parking space to a mini-park in 2005, complete with artificial turf, a potted tree, and a single bench (Figure 3.5). An early example of what is now referred to as "tactical urbanism," the guerrilla parking space park would only live for a few hours, as no permissions had been secured from the city. But a few hours were all that was needed to prove a point: As its creators observed from afar, the world's first "parklet" was soon in use, as passing pedestrians stopped to avail themselves of the small bench or to sit beneath the shade of the potted tree. Every year since, on global "Parking Day," this simple protest has been

Figure 3.5 First Parking Day installation at 1st & Mission Streets, San Francisco, November 16, 2005. Matthew Passmore.

reenacted in what now totals hundreds of cities worldwide, as planning students and community groups convert parking spaces to temporary parks, a transient but living simulation of the city that could be.

While there is no record of planning students jackhammering away the street on Parking Day, a growing number of municipal planning departments are doing so. Confronted with a rising epidemic of flooding events, ranging from "blue sky" inundation of streets in coastal cities (a flooding of streets from the daily high tide) to storms of unprecedented intensity, the extensive imperviousness of cities, particularly in the central business core, is a singular climatic stressor that falls fully within the purview of urban planners. For many cities, wide streets and a gridded moonscape of parking lots are the regrettable legacy of municipal laws from the mid-twentieth century requiring a minimum number of vehicle parking spaces for each new building constructed. In my own city of Atlanta, for example, many residential zoning districts continue to require two or more parking spaces for every house or apartment constructed, independent of the size or needs of the families who inhabit these homes. Similar parking minimum requirements are applied to commercial buildings based on their size and purpose. Implicated in hundreds of academic studies focused on a lack of affordable

Figure 3.6 Bioswale installation in New York. Richard Levine/Alamy Stock Photo.

housing and transportation in cities, parking minimums must also be understood as city-mandated formulas for flooding.

With more than 50 percent of many urban districts overlaid with impervious materials, and a sizable fraction of this total dedicated to the use and storage of private vehicles, roadway deconstruction is among the most effective strategies available to manage a rising frequency of flooding and drought in urban areas – and one that is rising in popularity. Less provocatively referred to as "green streets" programs in the United States, and more broadly as green infrastructure in the European Union, initiatives to deconstruct impervious paving in the form of on-street parking lanes, surface parking lots, and, in some instances, entire streets, are well underway. The specific rainwater collection and infiltration techniques being used include a wide array of vegetated trenches (bioswales) and engineered surface depressions for the storage and infiltration of rainwater (bioretention) and almost always entail the conversion of existing surface paving to create space for these systems (Figure 3.6). Evidence of a policy shift toward green infrastructure systems in cities can be found in the most recent US infrastructure law, which allocates funding to "reduce the extent of impervious surfaces" in cities as a means of better managing stormwater runoff – a sharp reversal from almost two centuries of federal infrastructure policy that helped finance the very paving soon to be uprooted [26].

A meaningful shift toward leveraging the surface area of cities to collect, store, and infiltrate rainwater, rather than diverting it downstream, represents for many cities a vast, untapped reservoir for shoring up drinking water supplies. The third component of SUDS entails the reuse of captured stormwater to either offset the quantity of municipally treated drinking water for outdoor watering or for purification and reuse by the city itself – an ultimate closing of the urban hydrologic loop that signals a radical reckoning with water scarcity in cities.

Showers to Flowers

Increasingly described as a war for water, intensifying disputes over the allocation of diminishing flows of the Colorado River in the American west may find their ultimate resolution in an earlier war fought on another continent. While the precise trigger for the Six-Day War of 1967 between Israel and a set of neighboring countries, including Egypt, Jordan, Lebanon, and Syria, remains to this day contentious, the centrality of water to this conflict is well established. Seeking to protect their own national water supplies, and, in the process, limit the flow of freshwater into the rapidly growing Jewish state to the south, Lebanon and Syria commenced water diversion projects in the mid-1960s to transport water from the Jordan River north of the Sea of Galilee, an important source of Israel's water supply at the time. The large and costly water diversion projects were never completed – a result, in part, of having been bombed by Israel – but the potential for water scarcity to sharply delimit the environmental and economic well-being of a growing national population in an arid region was made plain. Israel would not only establish a military buffer zone in capturing territory from Syria during the Six-Day War; it would also establish greater control over a critical watershed [27].

In the era of climate change, controlling a watershed may no longer prove sufficient to control one's supply of water. As the population of Israel doubled between the 1960s and 1980s, and then doubled again by the 2010s, the demand for drinking water has grown steadily. Israel has, somewhat famously, met this demand through the construction of five large desalinization plants along the coast of the Mediterranean Sea. The expense and energy intensity of these projects – with power mostly derived from the burning of fossil fuels – limit the output of these plants to drinking water only. To feed its growing population, Israel has also substantially increased its agricultural output over the last half-century, often cultivating arid desert soils for this purpose through the installation of extensive drip irrigation systems. Unable to draw upon desalinated water for this purpose, Israel would look to the same source tapped by water-limited Singapore: recycled wastewater.

Today, almost 90 percent of Israel's municipal wastewater is treated and piped to agricultural areas for use as irrigation water, achieving a national level of wastewater reuse three times higher than the next most proficient country (Spain reuses about 30 percent of its wastewater for irrigation) [28]. The result is an almost total reversal of the usual water collection and treatment process, in which freshwater is drawn from surface water or aquifers, used directly to grow crops or treated for municipal drinking water, and then returned to surface waters, finding its way eventually to the sea. In Israel, water is drawn from the sea, treated for drinking water, and then recycled for irrigation water, with any unevaporated residual returned to surface waters or aquifers. The result is the development of a system that is far less vulnerable to the vagaries of hostile neighbors and a changing climate and far more conserving of a limited water resource; but this approach comes with a high demand for energy.

To address this demand, Israel is seeking to trade water for energy with its neighbors. It is in a position to do so given the large volumes of seawater that it can desalinate and the almost total recycling of wastewater for agriculture. While the frequency and intensity of drought have been increasing throughout the Middle East over the last decade, Israel finds itself today with a surplus of drinking water. Confronting rapidly diminishing sources of surface and groundwater, Jordan has agreed to a deal with Israel through which it will construct and export 600 megawatts of solar photovoltaic energy – increasing Israel's annual supply of renewal energy by about 25 percent – in exchange for 200 million cubic meters of desalinated water, boosting the kingdom's supply of drinking water by about 40 percent [29]. The arrangement not only greatly lessens a growing water scarcity challenge in Jordan; it assists land-starved Israel in shifting from fossil sources of energy, which still account for more than 90 percent of its total energy production, to renewables [30] – lessening in the process the carbon intensity of its desalinization operations.

More than a simple water for energy transaction, the Israel–Jordan partnership will require an interconnection of water and power infrastructure, enhancing the resilience of both countries to address extreme weather events, cyberterrorism, and growing water scarcity. A parallel agreement increases the allotment of electricity to be delivered to the Palestinian Authority of the West Bank, ameliorating in some small part the generational injustices imposed upon the Palestinian population, which extend beyond basic sovereignty to access to critical resources. The deal falls well short of the least-first principle of climate adaptation but nonetheless highlights a key potential of the present moment: New alliances will be required to manage the ecological stresses imposed by an increasingly unstable climate. And these alliances carry the potential to enhance political stability in the bargain.

The extensive wastewater recycling that enables this scheme represents a new source of water available to cities confronting declining freshwater resources. It is an untapped wellspring that a growing number of cities are seeking exploit. Derided in an earlier era as "toilet to tap," a characterization intended to scuttle public acceptance, the recycling of wastewater for greywater or irrigation purposes is more enticingly branded today as "showers to flowers." As demonstrated by the Israeli experience, the reuse of wastewater for the purposes of irrigation enhances the volume of water available for both irrigation and drinking water, as treated municipal (or desalinated) water can be dedicated to potable end uses exclusively. In the North American context, where about a third of household water use is dedicated to the watering of lawns and gardens, the modification of municipal building codes to require or at least allow dual plumbing systems, through which water collected from sinks and showers can be minimally filtered by an in-home system and used to irrigate landscaping (or, at the very least, to flush toilets), would start to move cities in this direction. Water-starved Tucson, Arizona has mandated the installation of such dual plumbing systems for more than a decade; many if not most large US cities today prohibit the use of greywater.

Larger-scale applications of wastewater recycling repurpose the end product of municipal wastewater treatment facilities for municipal or agricultural uses rather than simply releasing this treated water downstream. In a growing number of towns and smaller cities experiencing prolonged drought conditions, city officials are offering deliveries of municipally treated wastewater to homeowners with storage tanks. Where agriculture is situated in close proximity to urban areas, there is potential for the adoption of the Israeli wastewater-to-food production model, but this model comes with a risk of crop contamination from both pharmaceutical and industrial residues in conventionally treated wastewater, including per- and polyfluoroalkyl substances (PFAS) – more commonly known as "forever chemicals" – that do not fully break down in the wastewater treatment process. The very real potential for this form of food contamination, and lack of near-term solutions, lends support to the reuse of conventionally treated wastewater for non-potable household purposes and landscape irrigation only.

The persistence of toxic substances in municipally treated wastewater elevates the need for a higher level of treatment, whether the treated water is returned to the watershed or recycled for use again within a city. With an ongoing shift to greater aridity in many parts of the world, the Singapore model of wastewater to drinking water is gaining converts. For more than a decade, one of the largest municipal water utilities in southern California has been treating a portion of its wastewater stream with reverse osmosis, harnessing a molecular level of filtration that yields virtually pure water. Remarkably, in the interest of side-stepping a public distaste for the direct recycling of waste-water, the Orange County Water District pumps this purified drinking water into a local aquifer, from which it is then ultimately pumped back into the conventional municipal drinking water treatment process and delivered to customers in a lower quality state than the purified wastewater injected into the ground. But this approach is soon to end.

As early as 2023, water utilities in the Los Angeles basin are planning to start augmenting conventionally treated drinking water with purified wastewater, a process known technically as direct potable reuse. Employing a set of filtration technologies similar to the desalination process, the City of Los Angeles is aiming to recycle 100 percent of its wastewater by 2035 [31]. A large fraction of this recycled water will be used as drinking water, boosting water supplies in drought-stricken southern California. To address the energy intensity of molecular-level water filtration, the region is also investing in dedicated renewable energy projects. If achieved, Los Angeles may rank as the first large city worldwide to close the urban hydrology loop and to do so without compromising an ambitious set of carbon reduction goals.

In the North American context, the deepest wellspring of new water is to be found not in its recycling but in a reassessment of its value. Copious amounts of treated drinking water are lost every day to leaks within the water delivery system and to a set of consumption patterns born of an underpriced resource. On average, 14 percent of treated water is lost to leaks in the United States (with a similar rate of loss in Canada), with some systems losing more than 50 percent of public water supplies to failing distribution systems [32]. Having in large part distributed the immense costs of constructing dams, large reservoirs, and inter-basin distribution networks across a national pool of taxpayers, the elevated costs of delivering drinking water in arid climates generally are not borne by those living there. Residents of rain-starved Nevada, for example, consume every year about double the quantity of municipally treated drinking water than do residents of Rhode Island, among the wettest states in the United States [33]. And residents of Rhode Island consume, on average, about double the quantity of residents of Europe [34].

Water conservation, the final and most maligned element of SUDS, is experiencing a resurgence in the era of Day Zero. Long the bane of resource economists, the appropriate pricing of drinking water to reflect its regional scarcity is a first and highly effective step in tamping down consumption. Not only do the most arid cities of the United States consume the most water per capita but residents of these cities also pay the lowest rates per gallon consumed [35]. While block rate pricing schemes, in which the price of municipal drinking water rises with higher rates of monthly consumption, have been adopted widely in North American cities and elsewhere, the graduated cost from one level to the next is often too low to much influence behavior. A key impediment to effective block rate pricing is that water utilities represent a critical source of municipal revenue; too much conservation can translate into budgetary shortfalls. As recounted at the start of this chapter, in confronting the rapid depletion of regional reservoirs, Cape Town revised block rate pricing schemes to inflict real economic pain on profligate users of water, with a comparable pricing scheme in the United States resulting in a monthly water bill of more than $2,000 – or about $4 per shower [36]. Importantly, even in the depths of the water crisis in Cape Town, the lowest-income households were provided a basic allotment of water at no cost.

The Cape Town experience with Day Zero is instructive, as few other municipal governments have adopted a comparable suite of aggressive water conservation strategies. Following the imposition of steep charges for excessive water consumption, city officials ultimately adopted a daily per capita water consumption maximum of 50 liters (~13 gallons). For households repeatedly violating this requirement, water management devices were installed to shut off water service once a daily threshold was reached – a restriction metered out to 46,000 high-consuming households. Dedicated units of the police were charged with patrolling neighborhoods for signs of outdoor water use and issuing tickets with large fines for lawn sprinklers, car washing, or the filling of pools. At one point, city officials harangued residents of noncompliant neighborhoods with a bullhorn, a very public doxing of those unwilling to acknowledge that the era of bountiful water had passed [37].

Perhaps the most aggressive tactic employed by the municipal government was the launching of an online household water consumption map, identifying properties that were meeting the mandated maximum monthly water limits with a green dot and, by the absence of a dot, those that were in violation of the water limits (Figure 3.7). Updated each month based on the prior month's metered water usage, the map represented an overt strategy to leverage public shaming to avert a situation in which all residents, wealthy or poor, would see their daily ration of water fall from 50 liters per person to half that amount, gatherable in buckets at public collection points. There is evidence that the online water map was effective, as the number of households complying with the usage restrictions increased by more than a third in the months following its deployment [38].

Several years removed from the crisis, Day Zero persists in Cape Town. With the enactment of an increasingly aggressive set of drought restrictions, Cape Town avoided a complete depletion of its drinking water system not through the tapping of new water sources but by stretching its limited reserves until the arrival of the summer rainy season, which delivered more rainfall than the two preceding years. At the height of the crisis, in the spring of 2018, Cape Town would succeed in reducing its daily water consumption by more than 50 percent – a feat of urban water conservation unmatched outside of shutting down the municipal water delivery system altogether [36]. Importantly, several years after the crisis, with reservoir levels mostly restored, Cape Town was consuming daily only two-thirds of its precrisis levels. For Capetonians, Day Zero is no longer a date on the calendar but a reconsidered mode of living in a water-limited world.

For many cities, urban water conservation is the most effective adaptation bridge available to manage accelerating water scarcity. While there are

Figure 3.7 Cape Town city water map. City of Cape Town, South Africa. Dark green dots identify parcels consuming 6,000 liters or less in the previous monthly billing cycle; light green dots identify parcels consuming 10,500 liters or less in the previous monthly billing cycle.

measurable cost savings to be realized from a shift away from water-intensive landscaping and lowered household consumption, significant investments are needed to replumb buildings for greywater recycling and, ideally, a capacity for rainwater harvesting. Residents and businesses will require monetary incentives for the installation of these systems, such as generous tax credits, coupled with revised building codes mandating their use, to bring about the near-term upgrades needed to lessen the demand for municipally treated drinking water. And municipal water utilities will need to more effectively set block rate pricing schemes to reflect growing water scarcity, rewarding low-consuming users with discounted costs while elevating the per unit costs of high-consuming users. Powerless to augment the annual receipt of rain, most cities have the capacity to make do with less – and to restore long-degraded urban watersheds in the process.

Radical Adaptation for Water Scarcity

The sustainable urban drainage system strategies explored in this chapter are highly effective in reducing the total annual demand for freshwater from rivers, lakes, and aquifers. Whether implemented individually or in concert, however, these techniques do not constitute a radical departure from conventional approaches to urban water management and will not be sufficient to sustain urban populations confronting steep reductions in regional water availability. Much like our approaches to managing extreme heat and flood risk, a radical approach to urban drought must concern itself with not only what strategies to adopt but how, where, and when those strategies are deployed.

As recounted in the two prior chapters, the first principle of radical adaptation is spatial decentralization. In the context of drought management, it will not be sufficient to construct ever larger drinking water reservoirs in a limited number of locations to enhance urban water storage – falling rates of rainfall and rising rates of evaporation are rapidly diminishing the utility of the large reservoirs already in place. Rather than enhancing the collection of a declining rainwater resource in a few locations we must enhance the collection of rainwater in all locations. A spatially dispersive approach to rainwater collection carries with it the benefit of exploiting the well-engineered capacity of the built environment, in the form of impervious rooftops and paved surfaces, to collect, channel, and, with only minimal modifications, store rainwater for future use or percolation to urban soils and groundwater systems. Rainwater collection in all locations further enables the harvesting and storage of water in the places where it is needed for toilet flushing, landscape irrigation, and the management of extreme heat, negating the need for conveyance systems to be constructed from centralized collection points.

Radical adaptation for drought adheres to the least-first principle in prioritizing the needs of frontline communities in the deployment of climate-adaptive infrastructure and services. It does so by recognizing that Day Zero is a meaningless milestone for households living in a constant state of water insecurity. The least-first principle requires that any new public infrastructure – including public drinking water collection points for informal settlements, wastewater management systems, and rainwater harvesting systems, as examples – be deployed in the near term to address the basic water needs of those lacking sufficient access to municipal water services outside of periods of drought. The least-first principle further requires that a basic minimum provision of municipal waters services – the delivery of drinking water and removal of wastewater – be provided to all households at low or no cost. Cities cannot effectively address a newly realized scarcity of water for most residents without first addressing the long-established scarcity of water for some residents. A radical approach to drought management sets a new minimum baseline for collective water security and then builds from this foundation. To be effective, drought adaptation must be both ecologically restorative and socially reparative.

The third principle of radical adaptation requires that municipal governments move outside of the conventional array of climate and environmental management strategies. In the context of drought management, a societal discomfort with the recycling of wastewater – a voluminous and valuable source of water available to all large cities – can no longer be accommodated in regions experiencing a pronounced reduction in annual rainfall. The recycling of wastewater requires not only that municipal governments seek to be innovative in the technologies adopted to enable this process but also that they must further be *nonnormative* in their long-established modes of operation. The redesign of cities to retain rather than discharge rainwater, to shift away from water-intensive landscape design, and to repurpose wastewater for irrigation, and, in some instances, drinking water, is to move outside of socially sanctioned norms of urban water management, at least in the context of highly industrialized cities. Necessitated by a delayed adaptive response to the now manifest stresses of climate change, radical adaptation will require that long-established norms of environmental management be reconsidered.

To be radical, urban adaptation to climate change must be dispersive, reparative, nonnormative, and, as explored in Chapter 4 on planned retreat, physically *deconstructive*. With rising seas regularly advancing the inland reach of the tides and the extremity of heat now surpassing the physiological limits of human tolerance in an expanding zone of almost every settled continent, some portion of many large cities and, in the most dire instances, the entirety of some cities, will not remain viable for dense human settlement. This is not a projection of apocalypse but an observation of efforts underway as I write to depopulate and resettle expansive areas of long-established cities. Largely limited at present to island nations, every low-lying city worldwide will confront the need to depopulate some portion of its land area; some cities in the most hot and humid zones of the planet may not remain habitable under even the most optimistic warming trajectories. In this sense, retreat is not a strategy but an eventuality. Planning for this retreat is the aim.

4

Retreat

As she crouched in the small dinghy enclosed within her neighbor's screened porch, eighty-four-year-old Marion Burkholder's most valued possession had been reduced to a disposable plastic cup. For her neighbor and copilot of the as-of-yet unlaunched lifeboat, the sole item hastily retrieved from her flooding home was a kitchen knife. As the storm surge from Hurricane Ian pulsed into the women's Cape Coral, Florida neighborhood on September 28, 2022 – twenty-four hours after a belated mandatory evacuation order had been issued – the plan was to ride out the rising waters in the small boat, using the cup to bale water and, if needed, the knife to cut through the porch's overhead screening. While the rising dinghy would not reach the level of the screening, it is not clear how the small craft with two elderly women would have fared in what was becoming the open sea. It could be observed that Marion lacked a good plan to weather a Category 4 hurricane, but that would miss a larger truth. Marion's plan, like that of so many Floridians, was to have no plan at all [1].

The no-plan-at-all plan is perhaps Florida's most lasting contribution to the realm of climate change policy. It is a plan that has played out over the course of Marion's life, commencing in the 1940s with the federally financed draining of wetlands across vast swaths of southern Florida. With the draining and filling of these wetlands, first for agriculture and later, more profitably, for expanding suburban development, Florida created an engine of booming real estate investment that shows no sign of slowing in response to a reversion of these engineered landscapes back to marshland and sea. The early successes in reclaiming land for orange groves around Lake Okeechobee in central Florida would pave a path to wetland conversion at the ocean's edge, where the many drainage canals needed to keep the constructed building sites dry could be marketed as waterfront suburban housing. By the time Marion purchased her home in the mid-1980s, Cape Coral could boast more mileage of canals than Venice, Italy – rising today to an astonishing 400 miles, each outfitted with suburban homes, pools, and docks but virtually no land protected from a hurricane of even moderate historical intensity (Figure 4.1).

Figure 4.1 Damaged structures along finger canals in Cape Coral, Florida in the aftermath of Hurricane Ian, October 2022. Adam Schultz/Alamy Stock Photo.

Despite this one strike against it, in the decade prior to Hurricane Ian, Cape Coral ranked among the top ten most rapidly growing urban areas in the United States [2].

More than a plan for responding to catastrophic weather, the no-plan-at-all plan is a blueprint for avoiding governmental expenditures for drinking water, wastewater treatment, public transit, and adequately funded school systems – all of which Cape Coral lacks in some respect to this day. It is a plan for shifting the costs of land development and continual redevelopment from the state to the federal government, helping enable in the exchange the elimination of state income taxes altogether (and drawing in more residents). The hubris of this lopsided arrangement reasonably can be characterized as breathtaking, but it cannot be said to be particularly risky. In the aftermath of Hurricane Ian, President Biden would fulfill the role of almost a century of predecessors in committing federal funds to the rebuilding of Florida coastal communities in the most hazardous hurricane zone found worldwide. Before the president's plane had lifted from Florida's water-logged soil in the days after Ian, a new tropical storm was spinning in the Gulf of Mexico.

The roots of a climate-catalyzed cycle of deluge and rebuild are to be found not with the first US settlement of Florida in the 1800s, nor with the massive water diversion and drainage projects of the mid-twentieth century. The modern blueprint for building in uninhabitable areas and shifting the recovery costs to others was conceived in the aftermath of Hurricane Andrew in 1992. The maiden named storm of the hurricane season that year, Hurricane Andrew was among the first of a rising frequency of mega-cyclones intensified by a rapidly warming ocean.[1] One of only four cyclones in US history to retain Category 5 strength at landfall – in this instance, just south of Miami – Andrew displaced 250,000 residents, claimed 61 lives, and caused about $50 billion in damages [3]. With more than 125,000 structures damaged or destroyed by 165-mile-per-hour winds, the hurricane also devastated the private home insurance industry in Florida, which was confronted with losses exceeding reserves. Unable to rely on private insurers to provide affordable home insurance policies to new residents in the aftermath of Andrew, a necessary condition for the securing of mortgages, the Florida legislature stepped in to create a new insurer of last resort, backed by the tax-financed capacity of the state government to absorb future losses.

With the establishment of Citizens Property Insurance Corporation, the no-plan-at-all plan would be given its modern form. In the absence of governmental policies mandating that property owners take steps to minimize losses from climate change, the private property insurance markets provide a critical economic signal of rising risk. To avoid a market-mandated retreat from the most vulnerable coastal areas of Florida, the publicly subsidized Citizens has continued to issue insurance policies long after a market case could be made for doing so. If insured losses from Hurricane Ian exceed reserves for Citizens, the state-chartered corporation has the power to assess fees on almost every insurance policy carried by residents in Florida (homeowners or vehicle insurance), regardless of location (coastal or inland) or which insurance company issued the policy [5]. Under the no-plan-at-all plan, the state effectively shifts the economic peril of living within the cone of next year's hurricane to those who live outside of it – keeping the cost of living in an intense hazard zone within reach.

More than a shell game for distributing the costs of hurricane recovery to noncoastal residents, the Citizens Property Insurance Corporation is a state-conceived vehicle for climate disinformation. Often deprived of the most basic

---

[1] The World Meteorological Organization manages a rotating list of hurricane names released in advance of the annual hurricane season each June. Names assigned to the most destructive storms, such as Hurricane Andrew, are removed from these lists due to historical significance. Hurricane Andrew was one of fifteen hurricane names retired in the decade of the 1990s, which was more than double the number of names retired in the 1980s [4].

scientific evidence pertaining to the pace and risks of climate change from governmental and popular media sources of information, residents of Florida must rely on the failsafe of steadily falling property values and rising insurance rates to signal growing risk. Citizens Property Insurance Company was conceived in the aftermath of Hurricane Andrew to disrupt this market signal of rising climate risk. When viewed in this light, the state's choice to pursue market manipulation in place of climate-adaptive building regulations makes sense: There is no potential to engineer resilience to 165-mile-per-hour winds and a 15-foot storm surge, at least not at a cost within reach of millions of snowbirds. To acknowledge the true economic risk of climate change in Florida is to acknowledge the necessity of a long-term deconstruction project and, with it, a rather inglorious end to Margaritaville.

Despite its roots in climate disinformation, the no-plan-at-all plan quietly acknowledges this one inarguable truth: There is no climate-adaptive building technology, green infrastructure strategy, or rapid evacuation plan that renders much of the Florida coastline viable for human settlement in a climate changed world. If the policy aims of the state are to be aligned with human welfare above short-term real estate dividends, there is only retreat. The central question for Florida moving forward, and for every provincial and national government worldwide, is whether the mode of retreat will be adopted before the next storm or after. The answer to this question, perhaps more than any other policy choice, will delineate the parameters of economic, cultural, and ecological resilience in the present century, as well as the shifting lines of Global North and South.

Retreat by Disaster

A little more than a decade after Hurricane Andrew in Florida, Hurricane Katrina would again raise the question of whether to retreat or rebuild in a major US city. The response in this instance would not take the form of an insurer of last resort but of a plan to depopulate the lowest lying areas of New Orleans. In the weeks after Hurricane Katrina breached New Orleans's levees and inundated more than 80 percent of the city's land area, a proposal to set aside some of the lowest elevation zones for floodable spaces – intended to lessen massive volumes of stormwater to be removed by the city's network of pumps – sought to address two principal concerns. First, reconstructed homes in these areas would be the first to flood again in response to the next storm or levee failure, placing the residents at greater risk of losses and committing the government to aid in a continual process of rebuilding. Second, setting aside

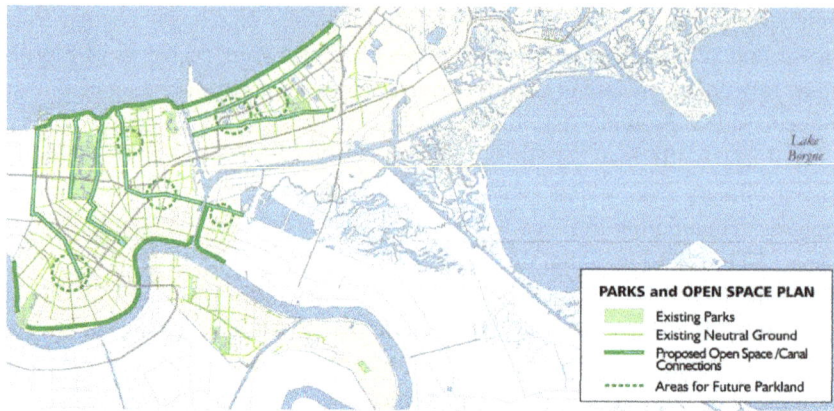

Figure 4.2 The "green dot" map from the Bring New Orleans Back Commission Action Plan. Bring New Orleans Back Commission, Urban Planning Committee, January 11, 2006.

flood control zones in several neighborhoods spanning the city from east to west represented (in the eyes of the plan's drafters) an equity-minded approach to enhancing flood resilience citywide.

Devised by a commission appointed by the mayor to direct the city's redevelopment efforts – the Bring New Orleans Back Commission – the proposal for new floodable parks was released to the public as a map with a series of green dots positioned over neighborhoods designated for the flood mitigation projects (Figure 4.2). While simple in design, encoded into this map was a policy choice for which there were virtually no precedents in the United States: In some neighborhoods, residents would be deprived of the right to rebuild on their own land. Under the prevailing regime of disaster recovery in the United States, not all residents are eligible for federal disaster assistance, but all landowners are entitled to rebuild on their own land, regardless of the number of prior disaster events or the risk of future flooding, wildfires, hurricanes, or any natural cause of property destruction. The Bring New Orleans Back Commission was proposing that the power of eminent domain be used to forcibly purchase land from homeowners unwilling to sell voluntarily.

Arguably more problematic than the proposed exercise of eminent domain was the siting of the green dots in the lowest-elevation neighborhoods only. As is the case in many noncoastal regions of the United States, to map the most flood-prone zones of the city is to map the zones most heavily populated by communities of color. Not a single green dot was located in a neighborhood in which a majority of residents was white. In New Orleans, elevation effectively equals race – a long-entrenched truism that should have been well understood by the mayor's commission. In a city in which 60 percent of the population is African American, a proposal calling for significant displacement in Black neighborhoods only carried little potential to accrue the needed political support for such an unprecedented action.

The Bring New Orleans Back Commission had broken the golden rule of community master planning: Never designate the *where* before amassing community support for the *what*. While many residents of New Orleans may support greenspace expansion to enhance flood resilience as a general idea, many fewer are willing to sell their land for this purpose – particularly in the emotional aftermath of a devastating event. Extensive greenspace acquisition in the period after Hurricane Katrina may have been possible but only through the identification of areas in which a large proportion of residents viewed the transaction as desirable. It is not enough to map where a planning intervention is physically needed; a city must also determine where the intervention is socially tenable – to do otherwise is to raise your opponent's army for them.

It is difficult to identify a major urban policy initiative with a shorter shelf life than the green dot map. In the lore of urban planning mismanagement, this episode seems destined to enjoy a more lasting run as a textbook example of policy malpractice than as an effective driver of neighborhood change. But to dismiss the green dot map on the grounds of its poorly executed rollout alone is to overlook its significance as perhaps the first attempt of a major US city to pursue large-scale urban deconstruction as a policy option for responding to climate change.

The green dot plan was not popular, but the selected policy of a right to return for all residents also can be viewed as an instance of governmental failure. In several of the neighborhoods targeted for flood adaptation, many years after the storm, less than half of the residents have returned, entire neighborhood blocks remain mostly vacant, and those who have remained or since moved in often find themselves more distant from schools, less well served by municipal services, and more lacking in community institutions than in other areas of New Orleans (Figure 4.3) [6].

Beyond a failure to well support residents of the neighborhoods most impacted by Katrina, the selected plan for rebuilding has also failed to enhance flood resilience for the city overall. As recounted in Chapter 2, New Orleans continues to regularly flood, most often in response to rainfall events that can no longer be characterized as extreme for the region. The gross inadequacy of municipal governance in the aftermath of Hurricane Katrina reverberates into

Figure 4.3 Vacant parcels in Lower 9th Ward, New Orleans, one of six neighborhoods designated by the Bring Back New Orleans Commission for the siting of
floodable parks. Google Maps, 2022, google.com/maps.

the present moment. While the absence of a plan to identify and acquire parcels
for a planned retreat program in advance of Katrina – a period in which
governmental acknowledgment of climate change at all levels was largely
absent – may have been defensible then, the persistent absence of such a plan
today is not.

The fundamental policy choice that confronts New Orleans and thousands of
other low-lying cities worldwide is not whether to retreat but how. With
a population standing at 80 percent of its pre-Katrina level, New Orleans has
experienced over the intervening years a profound retreat. But the mode of this
retreat has yielded virtually no benefits for the residents permitted to return to
the highest-risk flood zones nor to the surrounding neighborhoods. To depopulate areas of a city in response to the individual and uncoordinated decisions of
thousands of landowners, with no attempt to stage this depopulation in time or
to assemble contiguous parcels into hazard mitigation zones, is to adopt
a policy of *retreat by disaster*. Among the range of policy options for the
depopulation of areas confronting the greatest climate-related stresses, retreat
by disaster is the option most likely to carry the highest costs both for those who
elect to retreat and for those who elect to remain.

In the context of post-Katrina New Orleans, a retreat by disaster policy
would play out rapidly; in other coastal regions where gradual sea level rise
is the predominant climate stressor, the same policy unfolds more slowly.

If there is one unifying element to both of these scenarios it is an illusory absence of governmental agency. As a policy regime, the key feature of a retreat by disaster approach is that it posits each climate event to be an unforeseeable act of nature as opposed to an easily foreseen act of institutional neglect. Given that no one could have foreseen that the levees would fail, that a Category 4 storm would make landfall at the very site where coastal wetlands had been dismembered, that the electrical grid would cease to operate at the moment it was most critically needed (fill in the blank here), there is little governmental imperative to prepare for such improbable events in advance. Under this view, to develop a plan for retreat in advance of the next climate event is to acknowledge governmental culpability for climate change more generally, a road few elected officials have been willing to travel.

Local government officials reluctant to develop a planned program of retreat in advance of the next storm often cite national government policies for disaster recovery as an impediment to doing so. In post-Ian Florida, the US Federal Emergency Management Agency (FEMA) has provided two primary forms of disaster recovery aid to displaced residents. The first is the provision of temporary housing assistance in the form of payments for hotel rooms or mobile homes. The second is payments or loans up to $200,000 to enable homeowners to repair or rebuild homes anywhere in a federally declared disaster area, a massive zone stretching across twenty-seven counties and encompassing more than 50 percent of Florida's population. To be eligible for these payments, there is no requirement that a homeowner relocate outside of the flood hazard area or that a homeowner was carrying flood insurance at the time of Hurricane Ian (less than half of Florida homeowners living in FEMA-designated flood zones carry flood insurance) [7]. As noted, the number or extremity of prior flooding events, or projections of future flood risk, in no way limits eligibility for federal disaster assistance in the United States [8]. For one flood-prone property in Mississippi, FEMA's National Flood Insurance Program paid out thirty-four claims filed over thirty-two years, tallying a total expense about ten times greater than the property's value and possibly more construction time than required for the Great Pyramid of Giza [9].

In a world in which natural disasters were rare in any single location and delimited by the known extremes of a stable global climate, such a model of disaster recovery would be sound. If 500-year storms were to occur, on average, only once over many centuries, a national revolving fund for rebuilding could be fiscally managed such that payouts would almost never exceed reserves. In a world in which the return period for a natural disaster is unknown but grows shorter with each passing year – our present and now very long-term circumstance – such a model cannot be sustained. In this world, the economic costs of acquiring the rapidly growing inventory of properties at a high risk of storm damage is quite plausibly lower than the costs of a continual process of rebuilding – a policy option that lessens the risk of injury or death for residents and amasses valuable land for hazard mitigation.

If national government programs for disaster recovery are to become aligned with local government adaptation projects, such as floodable greenspace, shoreline buffers, or wildfire fuel breaks, the central thrust of these programs must shift from post-event payouts for rebuilding to pre-event compensation for relocation. In the US context, FEMA dedicates only a small fraction of disaster expenditures to the acquisition of properties most likely to be damaged or destroyed by extreme weather. Between 1989 and 2018 – representing most of the climate policy era – FEMA allocated about 1 percent of its disaster management funding to property acquisitions, with more than 99 percent of these expenditures dedicated to post-disaster emergency response and rebuilding in predominately high-risk zones [10,11].

While the US federal government has funded property buyouts in high-risk climate zones since the 1990s, the scope, voluntary nature, and highly bureaucratic administration of this effort provide only the most rudimentary basis for the scaling of a national program of planned retreat. With 90 percent of these property acquisition funds disbursed in the immediate aftermath of a federally declared disaster, less than 5,000 structures have been identified and acquired in advance of a high-probability climate event [9]. Put in context, in the more than thirty years since federal scientists first established the virtual certainty of property destruction in high-risk zones – and with an estimated 13 million residents now at risk of displacement this century [12] – the US government has succeeding in carrying out a planned retreat program for the rough equivalent of a single urban neighborhood.

A new approach is needed.

Retreat by Design

It is perhaps odd to observe that the first and, in some respects, most successful community-wide retreat programs in the United States were carried out well before the destructive potential of hurricanes would be amplified by climate change, and often with little to no involvement of the federal government. Flood plains, of course, will always eventually flood, and prior to the satellite era the precision in mapping flood plains across the Earth's land masses was imperfect, if such maps existed at all. Low-lying areas in proximity to rivers and shorelines that may have seemed ideal for settlement by a colonizing population with only limited knowledge of the hydrological rhythms of a region were often found later to be susceptible to periodic inundation. And the longer the return period between high water events, the more extensive a settlement was enabled to grow. More than a handful of towns over the nineteenth and twentieth centuries, confronted with repetitive and costly flooding events, would elect to relocate to higher ground [13].

One of the first documented community-wide retreat efforts in the United States took place in Niobrara, Nebraska, initially constructed on the banks of the Missouri River in the 1850s.[2] In the aftermath of an extensive flooding event in 1881, the roughly 500 residents elected to move the full town to a higher elevation site more than a mile from the river's edge [13]. To do so, numerous homes and other buildings were raised on jacks, lowered onto wheeled carts, and dragged by teams of horses to newly constructed foundations (Figure 4.4). That the town would undertake this lengthy and expensive relocation process again almost a century later, in response to flooding events in the 1970s, only burnishes its credentials as a successful model for planned retreat. In both instances, a recurring environmental hazard was acknowledged, a deliberative process was convened to identify a collective response, and a community, in its entirety, was deconstructed and reconstructed at a new, seemingly less hazardous site.

Arguably, a more instructive model for urban deconstruction is provided by one of the first urban-scale settlements on the North American continent. Situated just a few miles east of St. Louis, Missouri, several towering earthen mounds mark the site of a Mississippian settlement known as Cahokia. Established around 1000 CE and believed to have reached a population of 30,000–40,000 at its peak [15], Cahokia represents a remarkably large city and the economic and religious center of an extensive collection of settlements spanning much of the present-day southeastern to midwestern United States. Characterized by shared cultural practices, maize-based agriculture, and the linking of a network of villages by trading routes, Mississippian settlements also exhibited key elements of city-building, such as specialization around economic tasks and centralized management of community waste [16]. Given its size, Cahokia can be reasonably characterized, alongside Mexico's Teotihuacan, as one of North America's first cities.

The drivers of Cahokia's decline are not entirely understood, although an extended period of drought during the Little Ice Age of the 1500s generally is

---

[2] The name "Niobrara" is appropriated from the Omaha-Ponca language and is translated as "wide waters" [14].

Figure 4.4 Relocating a house to higher ground in Niobrara, Nebraska in 1881.
History Nebraska: https://history.nebraska.gov/wp-content/uploads/2022/10/46-
15-1.jpg.

believed to have played a role. Whatever the forces that led to the site's
abandonment, the city appears to have been designed for a regular process of
deconstruction. Across what must have numbered thousands of structures,
archaeologists find a consistent pattern of building demolition. First, wooden
poles composing the walls of houses were removed, followed by a filling of
postholes with variably pigmented clay, perhaps signifying a ritual purpose.
The floors of buildings were then burned in a fire consuming the possessions of
the building's occupants as well, such as tools or pottery. The purposeful nature
of these actions is attested to by the construction of new buildings directly atop
abandoned foundations, with structural timbers often driven into the same
postholes filled by the prior occupants. As observed by Newitz, "this ritual of
deliberate abandonment [sometimes] extended to entire neighborhoods ...

Perhaps, the Cahokians believed that every built environment had a set life-span, and always expected to close up the entire city one day" [17].

If there is a lesson to be distilled from prior instances of urban deconstruction in the recent or archaeological record, it may be this: To retreat in an uncoor-dinated fashion in the immediate aftermath of a disaster is not to retreat at all – it is to flee. The act of retreating, by contrast, is a deliberative and collective process, wherein a community elects through long-practiced custom or a political dialogue to relocate from a site of rising environmental (or other societal) risk to a site of lower risk. Importantly, such a deliberative process requires reliable information on the nature and timing of a hazard, and a collective framework through which partial or full deconstruction can be undertaken in partnership with government institutions.

Neither the recorded process of relocation undertaken at Niobrara nor the customary practice of building demolition revealed at Cahokia fully entails each of the elements required for a contemporary process of planned retreat. In relocating the town en masse in response to an anticipated future flood risk, the Niobrara model reflects a deliberative process and collective action yet does not give rise to an established protocol for future relocations or the setting of standards for the triggering of such a process. Likewise, the Cahokia model is one not of collective relocation but of a cultural predisposition to imperman-ency in the built environment (and, perhaps, in the making of a home). In this sense, a core value of the practice of building deconstruction at Cahokia is to be found in the routinization of such a process. In a climate changed world, urban deconstruction must transition from a mode of triage, as practiced at Niobrara, to a mode of regular and anticipated re-urbanization. It must be not only a practice but a shared value – just as growth has been the propelling value of recent industrial urbanization.

Retreat by design – a purposeful, routinized mode of collective urban deconstruction – can be understood as one of several modes of physical adaptation to climate change. Referred to in the technical literature as the "PARA" framework, climate adaptive responses by cities confronting present or projected impacts can fall into one of four general strategies: protect, accommodate, retreat, and avoid [18]. To *protect* a city from a climate-related impact, such as sea level rise, physical barriers or natural materials may be positioned between the rising seas and urban development, as is the case with seawalls or the nourishment of natural sand dunes. *Accommodation* strategies modify the built environment such that an intensifying climate impact can be managed without retreat. As explored in Chapter 2, in the context of sea level rise or other modes of flooding, accommodation strategies include the elevation of buildings or the creation of floodable greenspace within cities.

Avoidance strategies, in the technical parlance, pertain to development restrictions that prevent a city from encroaching further into a zone of increasing climate hazards and most often assume the form of zoning policies limiting new construction.

While the growing literature on climate adaptation presents several formulations of such adaptive responses, only three general classes of actions ultimately are available to us: modifying nature, modifying cities, or retreat. Of these responses, only retreat requires a decoupling of a community from the land it occupies and, as such, constitutes a mode of adaptation that is meaningfully distinct from other strategies. Venice with a deployable sea barrier and buildings retrofitted for wet floodproofing retains its essential identity as Venice by the act of *not moving* from its long-established physical location. The streets, canals, buildings, and piazzas of Venice represent more than physical infrastructure; these spaces constitute the physical setting of a shared history, with the collective memory of the city's lived events, over generations, encoded into the built environment itself. To move away from this fixed geography, even if the buildings were to come with us, is more than a departure from one urban location to another – it is a discontinuation of identity: communal, familial, individual. Venice cannot be made anew.

It is precisely this tethering of community identity to place that distinguishes retreat from other responses to climate change and renders it in practice not as an essential component of adaptation but as a final recourse once all other options have failed. But this characterization is largely an artifact of its poor deployment to date. When understood as one element in an interdependent framework for adaptation, retreat is not a threat to the life of a community but a central pillar of its continuation. When designed such that the displacement of those in the most hazardous zones of a city enhances the resilience of the larger number who remain, retreat becomes the catalyzing step for the array of protection, accommodation, and avoidance strategies required to mitigate climate impacts. If we are to have any success in managing the imminent and presently unfolding threat of climate change to cities, we must reconsider our most deeply held beliefs pertaining to the stasis of human settlement. With a reframing of these beliefs, what may seem radical becomes fundamental: *Retreat is the first step not the last.*

Evidence of a retreat-then-protect approach can be found in the limited number of cities, large and small, that have developed to date well-designed and ongoing climate adaptation programs. Among the first actions taken by New York in the aftermath of Hurricane Sandy was the acquisition of entire neighborhood blocks where flooding impacts were most concentrated. More than 200 houses along the coast in Staten Island, for example, were purchased with disaster recovery funds. In doing so, buyouts were prioritized for contiguous parcels of land that could then be combined into uninterrupted hazard mitigation zones. Once assembled, the US Army Corps of Engineers was then enabled to construct a new seawall and a series of wetlands at a farther remove from the water's edge, achieving a greater level of storm protection for remaining homes than would have been possible had a retreat process not been carried out [19]. The political viability of this approach is attested to by a recent bond referendum to enable further consolidated property buyouts and coastal flood protection projects, which was approved by more than two-thirds of New York voters in November 2022 [20].

A retreat-then-protect program is also underway in the small town of Paradise, California, which lost a majority of its homes and commercial buildings to the Camp Fire in 2018, among the most destructive and costly natural disasters in US history [21]. Similar to coastal flood protection, the creation of wide wildfire breaks, in which a corridor of land free of buildings or trees is established around a community, can greatly enhance resilience to future wildfire events. Since the 2018 fire, more than 300 acres of land have been assembled through a retreat program administered by the town's Recreation and Park District [22]. Importantly, the land acquired for fire breaks will not simply sit idle awaiting the next wildfire event but will be integrated into the town's network of parklands and designed for suitable recreation activities. Consistent with the Dutch concept of "room for the river," to be lasting, adaptation must serve the secondary purpose of public amenity.

In practice, the greatest challenge to a retreat-then-protect approach is in acquiring contiguous parcels of land through property buyouts. When structured as a voluntary program, as is the case in the New York and Paradise examples, a single property owner can prevent the amassing of contiguous land in the location where a hazard mitigation project may be most effective. In recognition of this complication, one provincial government in Canada has established a revised set of guidelines for property buyouts that seeks to balance the property rights of individual landowners with the collective interests (both fiscal and environmental) of the larger population. In the aftermath of 100-year flooding events in both 2017 and 2019, in which numerous homes in the City of Gatineau flooded repeatedly, the provincial government of Quebec revised its disaster recovery policies to offer buyouts or repair funds to homeowners in designated high-risk zones. For any property owners who accepted assistance in rebuilding, the government stipulated that no assistance would be available for future flooding events [18].

The establishment of a maximum limit on governmental disaster aid for any single parcel in a high-risk zone changes the calculus of retreat in important respects. No longer assured of governmental assistance for future flooding events, residents choosing to remain in high-risk flood zones confront the near-term effect of declining property values due to the unavailability of disaster aid for subsequent property owners. In direct contrast to Florida's climate management policies, Quebec's disaster relief program is designed to highlight rather than obscure the rising risk of climate impacts for residents. In doing so, the government has created a much stronger incentive for landowners to accept offers of buyout in the aftermath of a flooding event, lowering total governmental expenditures for disaster recovery over time (by eliminating liability for repeat flooding events) and better enabling the amassing of large, contiguous zones for hazard mitigation projects protective of the larger population.

In each of these examples, government-financed retreat is the first rather than the last step in a coordinated and continuing program of climate change adaptation. In each example, the inevitability of future flooding along the coasts of Staten Island or the banks of the Ottawa River, or the inevitability of future wildfire in the forest ranges of Northern California, mandates a governmental response that lowers population risks in the highest impact zones while enhancing community resilience for neighborhoods adjacent to these zones. In each instance, the scientific likelihood of a recurring climate-related disaster was assessed, a compensatory program of retreat was instituted, and various protection/ accommodation/avoidance strategies were implemented through a leveraging of the deconstructed land.

While none of these instances of effective climate adaptation resulted from a plan or framework established in advance of these climate events, from them (and others) we can distill four essential elements of a comprehensive program of planned retreat and adaptation: (1) map zones of high climate risk; (2) assemble land in high-risk zones through government-financed retreat; (3) mitigate climate risk for the remaining population through protect/accommodate/avoidance strategies; and (4) use the assembled land for community amenities. For simplicity, I refer to this approach as the Map–Assemble–Mitigate–Use or MAMU framework for adaptation.

Mapping of Climate Impact Zones. The defining element of a framework for planned retreat is a capacity to lower population risk and enhance community resilience in advance of a high-impact climate event. This will not be possible in all areas in which destructive, climate-amplified weather events unfold. Despite the great certainty of continued and accelerating planetary warming, and the array of catastrophic weather catalyzed by this warming, the precise timing and geography of many events remain too complex, too dependent upon the interplay of global, regional, and local phenomena, to predict with great precision. What can be predicted in advance with high scientific certainty is the likelihood of some impacts – in particular flooding – within a near-to-medium time horizon, such as a decade or two. Widely valued for its advances in the arenas of technology, medicine, and exploration (both past and future), the most essential dividend of science for managing climate change is often the least appreciated: Science buys us time.

While some degree of this dividend has been forfeited, in that the scientific case for pulling back from long stretches of the Florida coast, for example, or at least discontinuing development, met any reasonable threshold for risk assessment long ago, the tools for assessing risk in the interim have greatly improved, yielding a silver lining to our delay. These tools are in the form of climate simulation models operating at ever more resolved scales, remote sensing techniques enabling the precise measurement of the built and natural environments, and spatial analysis tools enabling the mapping of high-risk zones. Not only is it possible to accurately predict what spans of global coastlines will be inundated by rising seas over the coming decades but these maps are already developed and available to the almost 7 billion of us with smartphones.

Figure 4.5 illustrates one of these tools, the Sea Level Rise Viewer developed by the US National Oceanic and Atmospheric Administration and publicly available via the Internet. In response to a location query focused on any stretch of US coastline, the Sea Level Rise Viewer displays zones vulnerable to inundation from present-day high tides or future mean tide levels in response to projected sea level rise throughout this century. A complementary tool – the Sea Level Rise Projection Tool developed by NASA (www.sealevel.nasa.gov) – depicts similar information for coastal cities worldwide.

As depicted in Figure 4.5, extensive areas of Miami Beach, Florida are today regularly flooding from annual high tide events. Impacting a large number of cities globally, these "king tide" events can be pinpointed years in advance (the annual high tide is driven by the cyclical positioning of the Moon and Earth relative to the Sun) and can yield flooding of low-lying streets and buildings in the absence of storm activity. To find ourselves in a position where major global cities are routinely flooding without a single raindrop falling and with the precise date of the event known in advance is an outcome that ranks among the most profound failures of a scientifically advanced society. This is a failure not of science or technology but of governance.

We have at hand the technical information needed to identify high-risk zones for flooding from sea level rise over the near-to-medium term that is sufficiently precise to support planned retreat programs within these zones. To best enable such retreat programs, national governments will need to devise a set of inundation thresholds for classifying high-risk zones, develop protocols for

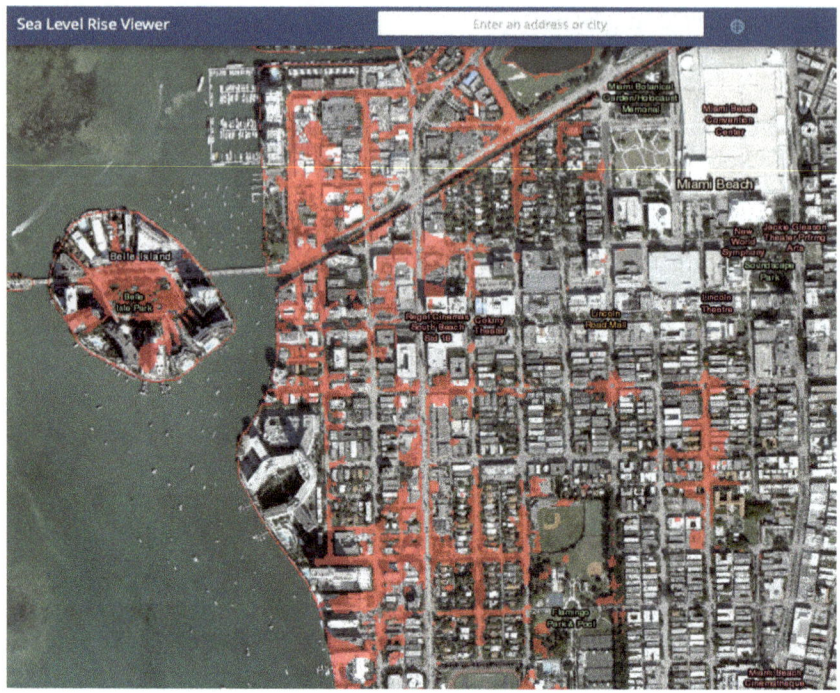

Figure 4.5 Present-day high tide flood risk in Miami Beach, Florida. Red areas denote zones subject to inundation from annual high tides. US National Oceanic and Atmospheric Administration, Sea Level Rise Viewer, www.coast.noaa .gov/slr.

the creation and regular updating of such flood zone maps, and share these resources with provincial and local governments for the administration of property acquisition efforts. In contrast to the green dot map of New Orleans, the needed maps must embody far more than the probable coordinates of future flooding events; these maps must embody a set of publicly sanctioned values differentiating what level of risk (to both individuals and communities) is tolerable from what level of risk is not. This is no small task.

For many national governments, a flood zone mapping framework predicated on statistical risk is already in place. What is needed is an updating of mapping protocols to reflect projected rather than historical sea levels and precipitation probabilities and a reorientation in risk management away from post-event rebuilding to pre-event relocation. Among the four dimensions of the MAMU framework, the mapping of climate impact zones is the only element with the groundwork largely in place.

Assembly of Contiguous Land. It is unclear if any nation worldwide has yet established an ongoing program for the acquisition of contiguous land in high-risk climate zones. While many national governments allocate funding for property buyouts in the aftermath of natural disasters, these programs tend to be limited in time to the period immediately following a destructive event and may lack incentives or requirements for the acquisition of contiguous land. The disaster assistance program established by Quebec to limit governmental liability for repeat flooding events, as an example, is only triggered by the instance of a flooding event, and so is not designed to proactively assemble land outside of these events. The Room for the River program of the Netherlands compelled the sale of property outside of particular flooding events, but this program was concluded in 2015 and is not an ongoing function of government. Likewise, the recently approved Environment Bond Act of New York allocates a limited quantity of funds for property buyouts over a fixed period of time.

A chief impediment to developing an ongoing program of planned retreat is that property acquisition as a governmental function tends to be institutionally relegated to the domain of emergency response operations. Designed more for the rapid clearing of land for rebuilding, as opposed to the enhancement of community resilience, the allocated resources and administrative expertise assigned to property acquisition programs are often poorly aligned with the imperatives of a planned retreat program. Framed more broadly, in most countries climate adaptation is not yet a recognized governmental function. There is no national agency of climate adaptation, dedicated budgetary allocation, or bureaucratic minister charged with a regular program of planned retreat as one of a portfolio of adaptation-related responsibilities. A simple expansion of the emergency response protocols in place in most countries will not provide an adequate basis for planned retreat because planned retreat, to be effective, must take place in advance of high-impact climate events.

The task of identifying privately owned parcels that are most well suited to acquisition and assembly into climate impact mitigation projects, working through a collectively sanctioned process to acquire those parcels, and assisting displaced individuals and families in lessening their own vulnerability through relocation constitutes one of the most ambitious governmental initiatives ever attempted. This is not a process that can be instituted episodically, nor can it be administered from the paramilitary footing of governmental agencies charged with disaster response. Planned retreat is a task for professional planners – those with formal training in risk assessment, community engagement, the many facets of land use management, and, not least of all, the heavy burden of environmental and racial justice. While different national systems of governance situate the planning function at differing levels of government, the need for climate adaptation to reside within agencies tasked with this function – or within newly created entities staffed by professional planners – is an essential first step for planned retreat.

In referring to the complex, emotionally charged, and institutionally precarious process of land acquisition by the single word of "assemble" – an admittedly technocratic and grossly oversimplifying mode of shorthand – it is not my intent to diminish the human experience of abandoning one's home, land, and, for most of us, the self-defining array of relationships tethered to place [23]. So profound is the distress of coping with displacement, or more generally the realization of how irrevocably we have already altered the ecological integrity of the places we inhabit, that an emerging concept in psychology, referred to as "solastalgia," seeks to formally characterize the experience and document the implications of this distress for mental well-being [24]. My assertion of the need for a routinized and continuing governmental function of land acquisition is intended not to overlook the reality of solastalgia but rather to anticipate and respond to it through the most robust and equitable process one can envision. If the need for retreat is unavoidable, what approach – by disaster or design – is most supportive of those to be displaced?

While the array of governmental tools for addressing the emotional distress of retreat is limited, the provision of new land for town or neighborhood-wide relocation offers one option that can enable the continuity of community ties. Among the few planned retreat efforts undertaken in the United States, most have entailed the relocation of an entire neighborhood or small community to a new tract of land situated at a higher elevation but within a limited radius from the historic location. One such project underway in Washington State is relocating more than 600 members of the Quinault Indian Nation situated at the mouth of the Quinault River and the Pacific Ocean, which is experiencing a rising frequency of flooding events, to a higher elevation site immediately adjacent to the present location. Assisted by a $25 million grant from the US Department of the Interior, the village has elected to move collectively in phases to the new site, through a process that affords the community full control over the design of the new settlement and the timing of retreat [25]. Both the proximity of the relocation site and the adherence to a community-driven process are intended to achieve the complete participation of residents and lessen the emotional distress of planned retreat [26].

Were the sole purpose of a planned retreat program to relocate vulnerable residents outside of high-impact climate zones, the need for full participation of all property owners within the zone would not be essential. However, as one of several steps in a larger framework for adaptation, land assembly carries the dual purpose of safeguarding those most vulnerable to climate impacts while enhancing resilience for the larger community. Given this, the success of a planned retreat program must be assessed with respect to the benefits imparted to those who are relocated and the benefits imparted to the larger community. If a refusal to participate in a property buyout program by a handful of landowners impedes the deployment of critical hazard mitigation strategies, such as the creation of floodable spaces or the installation of seawalls along coastlines, governments may seek to forcibly acquire private property through the legal process of eminent domain. This thorny topic brings us to the third element in the MAMU framework: mitigation.

Mitigation of Climate Hazards. Still in a formative stage, planned retreat efforts to date have largely focused on the relocation of small settlements – often Indigenous communities – confronting a gradual loss of long-settled or ancestral lands from sea level rise. For the much greater number of the more than 300 million worldwide that may need to retreat this century from coastal areas or inland flood zones [27], planned retreat will play out in larger urbanized areas. In the context of cities, the relocation of residents from high-impact zones yields critical space for the deployment of flood or other climate-related mitigation projects that would otherwise be difficult to site. This inherent value of land assembly is nonetheless often overlooked in post-disaster property buyout programs, which are largely designed to lessen governmental liability for future flooding events without much concern for adaptive resilience. The common outcome of this approach is a noncontiguous or checkerboard pattern of deconstruction, as illustrated in Figure 4.3, creating two significant burdens for municipal governments.

The first of these is the tremendous cost of maintaining basic urban services in underpopulated zones. Reliant on a fixed expanse of physical infrastructure in the form of streets, electrical transmission lines, and sewer and water systems, the declining number of houses that remain along a depopulating street require the same level of municipal investment while returning far less revenue in the form of property taxes. A long-standing problem in deindustrializing cities, the same economic considerations apply to densely settled neighborhoods threatened by sea level rise. Municipal governments and utility companies have a strong incentive for planned retreat programs to relocate entire neighborhoods, allowing these zones to be fully disconnected from costly infrastructure networks.

A second drawback of piecemeal land acquisition programs is the critical need to install new infrastructure – both green and grey – for the management of a rising climate threat to a growing proportion of the urban population. Adaptation projects designed to protect urban populations from rising sea levels, for example, will require extensive areas of land along coastlines, rivers, or other water bodies. Planned retreat efforts can supply land for this infrastructure in the very location it is most needed but only if all structures are removed from the site.

A complicating factor underlying most land acquisition efforts to date is the differing levels of government tasked with funding a retreat program and managing the land that remains. In the US context (as with many countries), national governments provide post-disaster assistance for property acquisition while local governments are given responsibility for managing the purchased land. Until recently, federal government agencies have sought to avoid forced property buyouts after storm events, and local governments have lacked the resources to implement their own retreat programs. This pattern appears to be changing, however, as the US Army Corps of Engineers – the federal agency charged with developing flood management infrastructure – quietly modified its flood mitigation policy in 2015 to require municipalities to compel property sales through eminent domain if needed to complete a project using federal funding [28]. More recently, in 2022, the US Government Accountability Office identified the use of eminent domain as one of six potential policy changes that could improve FEMA property acquisition programs [11].[3]

For governments uncomfortable with the use of eminent domain for the acquisition of property in high-impact climate zones, the approach to disaster assistance recently adopted by the Canadian province of Quebec – let's call it the 'Quebec compromise' – may offer an effective middle way. The establishment of a parcel-specific limit on disaster assistance that governs in perpetuity public expenditures on rebuilding provides a strong incentive for property owners to accept offers of a buyout without legally compelling a sale. For property owners declining assistance in relocating, subsequent disaster events – all the more likely in designated high-risk zones – may ultimately compel a move in the absence of governmental support, achieving with more time the same outcome as mandatory property acquisition.

Coupled with resources to support property acquisition programs, whether voluntary or compelled, is the critical need for governments to assist individuals and families occupying rental property within zones targeted for planned retreat. Short-term rental assistance is commonly available in the aftermath of disaster events, but these resources typically are not designed for residents displaced by planned retreat programs. With the expectation that rental housing within areas designated for retreat and adaptive infrastructure will never be

---

[3] The US Federal Emergency Management Agency (FEMA) prohibits municipal and county governments, which are charged with administering federally funded property buyout programs, from compelling the sale of land with the power of eminent domain.

rebuilt, municipal governments will need to further invest in the stock of affordable housing outside of these zones – an acute need in the present moment that will only be intensified by a scaling-up of retreat efforts.

As discussed in Chapter 2, climate adaptation and affordable housing must be understood as variants of the same general municipal obligation; more vulnerable to the threat of climate change than those housed in high-impact zones are those who lack stable housing altogether. For some cities, opportunities for affordability may be found from advancing housing into the water bodies themselves, such as is underway in the Netherlands, if such communities can be engineered to withstand extreme events. In other cities, higher density construction in low-risk zones and a more robust approach to rent stabilization will be needed to expand the availability of climate-resilient and affordable housing.

Importantly, the entwined challenges of climate resilience and affordable housing will not play out in coastal cities alone. Most large cities worldwide are positioned along large bodies of water, including rivers and lakes, and virtually all large cities encompass low-elevation zones subject to flooding from intense rain events. In other regions, adaptive infrastructure may be needed to manage the rising threat of wildfire or drought, wherein expanded greenspace is needed for groundwater recharge and storage. While the scale of retreat required for inland cities is likely lower than mandated by the rising seas, all cities can enhance adaptive resilience through deconstruction within the most high-risk zones. Combined with the need for a large-scale resurfacing of cities to lessen heat exposure and stormwater volumes, the de-urbanization of some zones within cities creates an opportunity for a re-urbanization of others. As cities are redesigned for climate resilience – likely the central aim of the urban planning profession over the coming decades – affordability and livability must be both the object and the method of transformation.

Use of Land for Community Amenities. Situated only 30 miles from Cape Coral and exposed to virtually the same windspeeds from Hurricane Ian, Punta Gorda, Florida would experience far less damage to homes and commercial structures than its neighbor to the south. At first glance, there is little physically to distinguish the two communities. Both are perched along river inlets from the Gulf of Mexico; both are protected by barrier islands from the full force of hurricanes; and both are largely fashioned from finger canals carved from coastal marshlands. But while one would experience some level of damage to virtually every building in the city, the other would see only 5 percent of its structures require significant repairs [29,30]. One explanation for the different outcomes can be credited to building codes: In the aftermath of Hurricane

Charley in 2004, which resulted in tens of millions of dollars in property damage for Punta Gorda, municipal building codes were revised to require much higher standards for wind resistance. Having suffered more damage from Hurricane Charley than other areas, a larger percentage of homes in Punta Gorda were rebuilt to weather higher windspeeds. A second change in the aftermath of Hurricane Charley would further differentiate Punta Gorda from its neighbors: the adoption of a climate adaptation plan.

Such plans, of course, are only as effective as the municipal actions that follow from them. But to even have such a plan on the books in 2009 placed Punta Gorda in rare company, both within and outside the State of Florida. More impressive, the plan initiated a program of planned retreat that would result in the relocation of a municipal building and establish a process for the acquisition of homes in the most vulnerable areas. Punta Gorda from its founding has prioritized waterfront land for public parks, leaving little room for the construction of houses along the shoreline. On the eve of Hurricane Ian, hundreds upon hundreds of houses sat immediately adjacent to coastal waterways in Cape Coral. In Punta Gorda, the number of houses immediately adjacent to coastal waterways numbered less than forty.

There is little doubt that the societal cost of planned retreat in any location – borne by both individuals and governments – will be far lower over time than a cycle of repetitive rebuilding. This cost will be, nonetheless, a substantial sum. In deciding for so long not to make a decision on climate change, we have elected a path of maximal costs – a path that now requires a radical mode of adaptation relative to the policy options available to us several decades ago when the science attained high confidence. Retreat is now, in one form or another, underway upon every continent, and it carries with it the potential to work against both the effects and the drivers of climate change.

Planned retreat works against the effects of climate change by enhancing the resilience of cities to moderate the impacts of storm surge, wildfire spread, groundwater depletion, and intensifying heat waves. The success in sustaining one inhabited zone of a city by transforming another into adaptive infrastructure is the principal return on the public investment in property buyouts. A secondary return can be achieved in the form of expanded public greenspace and networks for human-scaled modes of transportation – a set of strategies that can assist in the needed transition away from auto-dependency.

It is perhaps surprising to find so little emphasis on the automobile in a book about climate change. Responsible for almost a quarter of carbon emissions in the United States [31] – a fraction that may be at long last peaking with the scaling of electric vehicle (EV) production – vehicles propelled by internal combustion engines are a key focus of climate mitigation programs. Often lost in the discourse over the EV revolution is the heavy toll cars of any propulsive technology have imposed on urban environments and our experience of them. In the US context, more than half of all land in downtown districts often is dedicated to the movement and storage of personal vehicles [32]. When considered as a proportion of urban public space, we have handed over the vast majority of our collectively held land to cars. Such a disproportionate allocation of land to privately operated vehicles plays a central role in the present housing affordability crisis, and it impounds urban acreage critically needed for heat, flood, and drought management.

The final step in the MAMU framework entails the opening up of assembled land to public use. In the small city of Punta Gorda, this use takes the form of a park and bike trail network anchored to protected land along the coastal waterway and running alongside a protective seawall. The network of dedicated bike paths spans 18 miles and provides a viable alternative to vehicle travel for most neighborhoods in the city. In larger cities, such as Nijmegen in the Netherlands, planned retreat projects are contributing to alternative transportation networks more heavily used than the roadway system. As part of the Room for the River program, numerous residential structures were acquired to expand the floodway of the River Waal, adding 618 acres to the city's public lands and expanding the network of biking "superhighways," including a new bridge spanning the river (Figure 4.6). Far more than an amenity for weekend touring, the regional network of dedicated bike trails provides priority to cyclists at intersections, connects regional cities and towns, and, within Nijmegen, accounts for two-thirds of all commuting traffic [33]. When leveraged for alternative transport infrastructure, planned retreat carries both adaptive and mitigative benefits.

The tension between the vast acreage of land dedicated to vehicle transport and the land needed for climate adaptation will extend beyond projects of planned retreat. The first principle of radical adaptation – a dispersive approach to infrastructure siting – requires that greenspace for heat and flood mitigation be set aside throughout the full extent of cities, not simply along the edges of coastlines and waterways. The conversion of on-street parking and surface parking lots, in whole or in part, to green infrastructure in the form of bioswales and tree canopy provides a highly effective approach to climate adaptation that can be replicated uniformly across the auto-dominated landscapes of cities. With a reduction in the land available for the movement and transport of personal vehicles, infrastructure for alternative modes of transport – walking, cycling, and public transit – can be integrated compatibly into newly deconstructed spaces with seawalls, floodable parks, wildfire breaks, or urban micro-forests. The same is not true for a replacement fleet of electric vehicles. In a climate changed world, theories of ecology and

Figure 4.6 New cycling bridge constructed as part of a planned retreat project in Nijmegen, Netherlands. Laurence Delderfield/Alamy Stock Photo.

economics are converging: Endless lanes of on-street parking are no longer the highest and best use for scarce urban space.

Making use of repurposed land in cities for non-vehicle transport networks, expanded parkland, performance space, public art, recreation facilities, and restored habitat, alongside climate adaptive infrastructure, transforms our cities into environments that are more livable and beautiful while enhancing community resilience to climate change. Both the livability and the affordability aims of urban planning that are already bringing about significant changes in many cities today can also be leveraged for the purposes of climate adaptability – but only if these aims are understood to be highly codependent. To this end, planned retreat is not a failure of municipal management but one step in a too-long delayed process of re-urbanization – a process that can yield dividends much broader than climate resilience alone.

Retreat As Radical Adaptation

My purpose in this chapter has been to propose a general framework for planned retreat that embraces the process of urban deconstruction as both an imperative and an opportunity. I do so aware of the uncomfortable pairing of these two words – "urban" and "deconstruction" – as this phrase is not a part of the planning lexicon. But it should be, and not for the purposes of climate adaptation alone. All cities will eventually confront the need to restructure the built environment in response to changing ecological, economic, or social conditions. In prior cultures, the ultimate need for deconstruction was in some instances designed into the structure of buildings or urban blocks in the nascent stages of city making. Of all the patterns of urbanization to follow, the postindustrial, auto-oriented design that characterizes much of the metropolitan landscape of North American cities today, and that which regrettably has been exported to many countries of the Global South, is perhaps the least adaptable to the physical retrenchment and repurposing now necessitated by a rapidly changing climate.

In my own city of Atlanta, hundreds of miles from the coast, retreat is a regular process of urban change. Numerous auto-oriented commercial districts of the mid-twentieth century, followed by the first wave of big-box retail developments and a wide range of industrial sites, sit shuttered and disconnected from surrounding neighborhoods by expansive moats of surface parking. No longer contributing to the economic viability of the region, absent the occasional dystopian film production, these sites persist in an indefinite state of blight that can stretch for decades – a mode of postindustrial retreat that yields no adaptive benefits for the adjacent neighborhoods. To the contrary, these sites continue to generate excess heat and vast quantities of stormwater, sustaining the collective costs of these developments long after a stream of property tax revenue has ceased to flow. No large US city lacks these derelict sites; no city would fail to benefit from their disassembly.

Urban deconstruction, as both a cultural value and a bundled obligation of land ownership, anticipates the need to return the built environment to a natural state once a socially productive activity has discontinued or once an environmental hazard exceeds a collectively sanctioned threshold. Consistent with the principles of radical adaptation introduced in prior chapters, retreat is only perceived as radical due to its omission from conventional modes of urban planning. Flush with an almost limitless expanse of appropriated land and fueled by an almost limitless cache of cheap energy, the urban planning profession of the last two centuries was chartered for the purposes of directing and sustaining the physical growth of cities; the need for degrowth has been, until the very recent era, almost entirely unconsidered. The necessity of a regular process of deconstruction is today made obvious by the rising seas, but its logic and social utility have been apparent from the earliest of urban settlements.

In the context of climate policy, planned retreat has been characterized as one of several tools for adaptation, and a tool most often deployed last in a sequence of actions. It has been my aim in this chapter to suggest that retreat is not simply one of several arrows in the adaptation quiver but rather the fabric of the quiver itself. Planned retreat enables other adaptive responses through its assembly of the land needed to carry out programs of protection, accommodation, and avoidance most effectively. Adapting cities to the threats posed by heat, flooding, drought, wildfire, and emerging hazards not fully anticipated requires extensive investment in protective infrastructure – both green and grey – and this infrastructure requires large assemblages of land in the highest risk zones. The establishment and maintenance of this infrastructure govern the nature and extent of the accommodation strategies that will be required in areas adjacent to zones of retreat. And the land assembled for retreat constitutes the core areas to be protected through avoidance programs. As such, planned retreat is not the last step in a program of climate adaptation; it is the initial, enabling step for what follows.

Of the four principles of radical adaptation I introduce in this book – supportive of an adaptive urbanization that is *dispersive, reparative, nonnormative,* and *deconstructive* – only the last of these, planned retreat, requires a reconceptualization of the process of urbanization itself. Neither the decentralization of systems for environmental management, the prioritization of adaptive investment toward the generationally marginalized, nor the introduction of novel modes of living requires that the process of urbanization be understood for the first time in de-expansionary terms. In a climate changed world, the generally accepted preconditions for urban settlement – a minimum population size, a system of governance, modes of economic and cultural production, systems of environmental management, and, more recently, a sustained process of physical growth – must be broadened to address the long-neglected imperative of ecological integrity.

Through retreat, we can make room for the river, yes – but we can also make room for a broader ecology that has ever quietly and invisibly sustained human settlement to this moment: water, sunlight, plant life, the recycling of nutrients, a greater symbiosis with other species. These are the building blocks of adaptation. This is neither a new nor a radical idea.

Postscript: Vine City

In the summer of 2016, I was invited by a colleague at the US Environmental Protection Agency to participate in a novel experiment. Federal funding was being provided to transform a large, multilane road in a westside Atlanta neighborhood into the city's first "green street." The roadway redesign project would entail the conversion of a lane of traffic into a running series of bioswales, vegetated spillways, and numerous new street tree plantings. More than a stormwater management project, the transformation of a half-mile stretch of roadway would also have the effect of increasing canopy cover along a street almost entirely denuded of vegetation, with likely benefits for heat management. The proposal was to install temperature sensors along the roadway a year in advance of the green street conversion to measure the effects of a natural experiment in heat adaptation. I readily accepted.

The green street project was just one component of a larger plan to manage flood risk in what may be Atlanta's most climate-vulnerable neighborhood: Vine City. Situated on the western edge of the city's highly impervious downtown district, and at a lower elevation, the Vine City neighborhood is subject to substantial flows of stormwater runoff during rainfall events. In one particularly heavy event in the summer of 2002, as the remnants of Tropical Storm Hanna brought more than 15 inches of rain to Georgia, rapidly accumulating volumes of stormwater would soon overwhelm the neighborhood's antiquated combined sewer system, leading to a flooding first of streets and then the front yards of residents' modest homes. But the most acute flood hazard would come not from without but from within, as geyser-like effusions of combined stormwater and raw sewage literally blew the lids off of toilets and gushed from sinks. Rising to a height of more than six feet within some homes, the putrid floodwater foreclosed any option for rebuilding across several neighborhood blocks [1]. In Vine City, retreat was underway.

In the aftermath of the flooding, the city offered to purchase more than fifty homes and to assist in relocating residents. More than a decade later, this assembled land would provide the space needed for the construction of

Figure PS.1 Rodney Cook Sr. Park in Vine City, Atlanta. Georgia Flash/Alamy Stock Photo.

a large floodable park capable of storing and infiltrating more than 10 million gallons of water and greatly enhancing the flood resilience of the surrounding neighborhoods (Figure PS.1). Running alongside this park would be the city's first green street, representing in combination Atlanta's largest green infrastructure project to date.

While many cities worldwide are moving forward with climate adaptation efforts, some quite aggressively, virtually no city is pursuing a comprehensive approach well aligned with the principles of radical adaptation presented in this book. With no formally developed climate adaptation plan in hand or, as I write, in development, Atlanta, Georgia is precariously unprepared for the range of climate impacts unfolding in the present moment. But in the small neighborhood of Vine City, through the efforts of several government agencies, foundations, and community organizations operating independently and in concert, a radical approach to managing climate change is beginning to take shape. There is no more worthy place for it. In Vine City, the threads of an adaptive framework that aims to be dispersive, reparative, nonnormative, and deconstructive can be observed, demonstrating the potential of an approach that emphasizes multiple dimensions of community resilience.

A slow-moving process of retreat was underway in Vine City well before the 2002 flood. Initially settled in the late nineteenth century, the community stretching in two directions from the centrally positioned Vine Street offered one of the few areas open to African American home ownership in the decades to follow. During this period, Vine City and adjacent neighborhoods would emerge as one of the most politically and culturally significant centers of African American life in the United States. Home to the prestigious historically Black institutions of Morehouse and Spelman Colleges and numerous leaders of the civil rights movement, including Dr. Martin Luther King Jr. and his family, the area offered a stable middle-class community to Black Atlantans denied access to other neighborhoods. Yet with the systematic disinvestment brought about through redlining, the displacement of residents through nearby interstate highway development, and mid-century urban renewal projects, Vine City entered a period of sustained decline. By the period of the Great Recession, in the late 2000s, more than a third of the neighborhood's parcels stood vacant, a quarter of all residents lived in poverty, and the neighborhood experienced more violent crime than almost any other nationwide [2,3].

That the largest green infrastructure project undertaken to date in Atlanta is located in one of its most distressed neighborhoods distinguishes the city's early forays into climate adaptation from other regions. As noted in Chapter 2, a series of high-profile adaptation projects in New York and New Orleans have disproportionately benefited middle-to-high-income neighborhoods. A study of recently established resilience districts across global cities includes not one located in a low-income or otherwise disadvantaged neighborhood [4]. The siting of major climate adaptation infrastructure in a small neighborhood that exhibits among the highest poverty and land vacancy rates citywide attests to a reparative approach in the prioritization of adaptation investments – an approach that extends beyond flood resilience alone.

In concert with climate adaptive infrastructure, a partnership between city agencies and local foundations is expanding affordable housing in Vine City and adjacent neighborhoods. While the availability of housing within reach of typical neighborhood incomes has been falling across Atlanta, in Vine City the stock of affordable housing is growing. In addition to the construction of new affordable housing, the Westside Future Fund – one of several organizations focused on community resilience in Vine City – preferentially leases new units to existing residents of target areas, creating a counterweight to rapid gentrification of historic westside Atlanta neighborhoods. Additional tools to expand affordable housing and retain long-term residents include assistance with down payments for single-family homes and annual payments to offset the rise in property taxes for legacy homeowners [5].

Other programs designed to enhance community resilience – social and economic in addition to environmental – include new schools and community centers for youth, workforce training initiatives, and a pilot program offering new houses at a reduced cost plus a monthly stipend to Atlanta police officers who commit to living in the neighborhood for five years. The early returns on these programs and others are promising: metrics for reading, math, and science at a new elementary school have increased in recent years by 300 percent; more than 700 residents have found jobs in the community through workforce development programs; violent crime rates have fallen by more than 40 percent; and rates of land vacancy and unemployment in Vine City have fallen over the last decade at a higher rate than adjacent neighborhoods [2,6]. The investment in these various community programs and institutions, in concert with new green spaces, is substantial – in excess of $50 million of public and foundation funds – but so too is the return on investment for a long-neglected community. And valuable lessons are being learned by the city agencies and nonprofit organizations working in Vine City.

One of these lessons is that an array of conventional poverty alleviation programs implemented over decades had not yielded measurable results in westside Atlanta neighborhoods. What appears to be working in recent years is a set of nonnormative approaches designed in direct collaboration with community residents. One of those novel ideas is that the governmental and institutional actors working in Vine City must physically ingrain themselves within the community their programs are seeking to impact. The provision of affordable housing to police willing to reside in the neighborhood and the opening of branch offices of several foundations working in Vine City have brought for the first time a continuing and sustained presence of institutional actors – described by the vice president of one foundation as "nodes of stability" [6]. Another community request was that an investment be made in public art. More than forty, often large murals and street art displays have been commissioned in Vine City and adjacent neighborhoods – a modest but meaningful investment in community identity (Figure PS.2).

Among the most prioritized goals expressed by neighborhood residents is that the wealth of vacant land resulting from years of flooding and property abandonment be repurposed for community use [7]. One result of this goal, and of the efforts of numerous community organizations, is the cultivation of a more extensive number and size of urban farming plots in and around Vine City than anywhere else in Atlanta. To walk in any direction from the center of Vine City is to eventually stroll past one of at least ten active farming sites ranging in size from an eighth of an acre to almost four acres. The largest of these – the Truly Living Well Center for Natural Urban Agriculture – grows

Figure PS.2 Mural at 320 Sunset Avenue by Muhammad Yungai. Photo by author.

more than 30,000 pounds of organic produce a year, in addition to producing honey and tilapia on site. Part of a citywide effort to address food deserts in Atlanta, the many community gardens and urban farms now active in and around Vine City have helped increase the number of residents located within a half-mile of fresh produce growers or grocers from 52 percent to 75 percent over a five-year period [8]. The urban art and agriculture efforts underway in these historically marginalized neighborhoods represent a creative set of tools for enhancing adaptive resilience through a leveraging of the capacity, resources, and desires of the community itself.

Further along than any other neighborhood in Atlanta and, quite possibly, the country as a whole, Vine City is constructing an integrated network of floodable parks and other green infrastructure elements designed to restore the hydrological function of a riparian wetland that was never well suited for urban development. Located downslope from the headwaters of the Proctor Creek watershed, the neighborhood occupies one of several bottomland areas across Atlanta initially made available to African American settlement due to its recognized propensity for flooding. Exacerbated over time by the increasing imperviousness of the surrounding neighborhoods and, more recently, a rising intensity of rainfall with climate change, what was once an irregular pattern of high-water events has grown more frequent. A restoration of the underlying hydrology represents not only the most effective approach to managing stormwater in Vine City but the least costly alternative. And it carries with it a set of neighborhood dividends long lost to a network of buried streams and

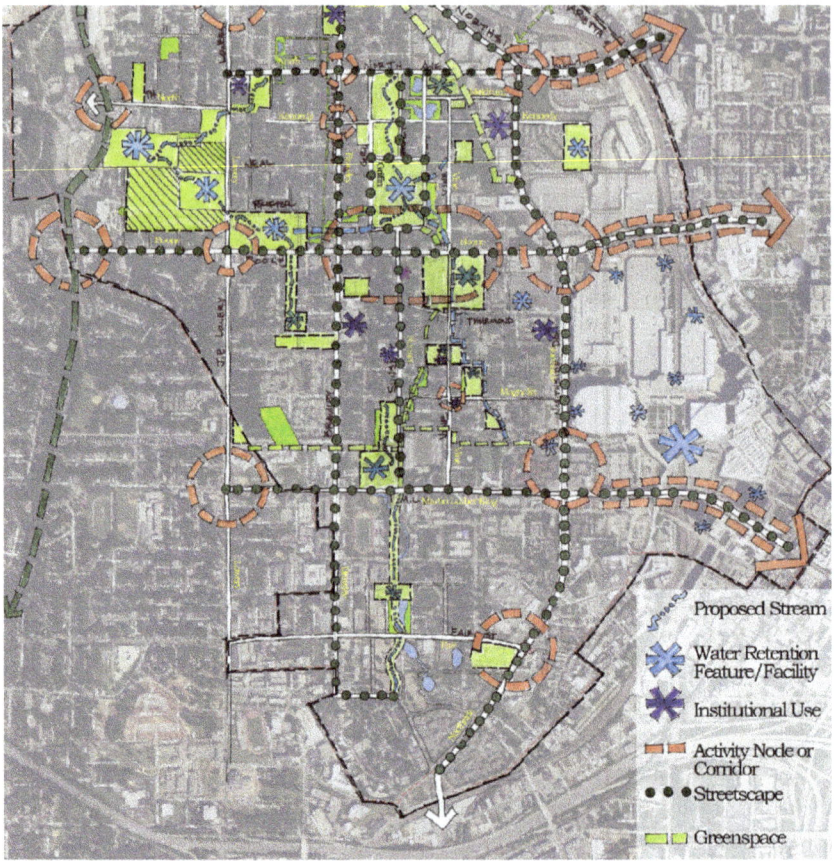

Figure PS.3 Green infrastructure network developed through Proctor Creek North Avenue Watershed Basin plan. Park Pride, 2010 [7].

stormwater pipes: space for recreation, space for food cultivation, heat management, wildlife habitat, and connectivity to a citywide network for non-vehicle transportation.

The blueprint for this ecosystem restoration is a plan developed more than a decade ago by Park Pride – an Atlanta-based organization focused on the expansion of urban greenspace – in collaboration with community residents (Figure PS.3) [7]. While the centerpiece of this plan would become the large floodable park assembled from the 2002 land acquisitions, its key innovation is to be found in its dispersive design: More than twenty new and restored greenspaces, connected by daylighted stream segments and a network of green streets, are providing far more stormwater management capacity than would be possible with a single, centralized greenspace alone. To date, four of the envisioned floodable parks have been completed, providing a combined capacity to manage more than 14 million gallons of stormwater and spatially distributing the infiltration and conveyance of this runoff to natural waterways. Few neighborhoods worldwide can claim the same density and connectivity of green infrastructure as Vine City – a statistic rendered more remarkable by the marginalized history of the community.

To be sure, Vine City remains among Atlanta's neighborhoods most vulnerable to climate change. Despite the ongoing expansion of greenspace, the neighborhood continues to lag behind most other residential areas in the extent of tree canopy; its proximity to the heart of the city's heat island in the downtown district exposes residents to a greater intensity of heat than most other residential zones. On the whole, residents are poorly prepared to cope with an elevated heat burden. More households lack air conditioning than almost every other neighborhood in Atlanta, and households experiencing energy scarcity are less likely to use mechanical cooling systems if available. The persistence of economic distress among residents has the effect of amplifying all climate stressors: extreme heat, flooding, drought, and a rising trend toward infrastructure failure. The ambition of climate resilience in Vine City remains, without a doubt, a long-term and uncertain project.

But it is at least a project underway. In sharp contrast to most urban neighborhoods worldwide, Vine City is less flood-prone this year than last; less neglected by the governing institutions charged with enhancing environmental and economic resilience; and, if the available statistics are to be believed, less impoverished overall. Vine City has commenced a community-driven effort to work against the extremity of its ecological degradation. This is a remarkable transition in a relatively short period of time and, if we allow it, more than a little hopeful.

In closing, I will return to the idea at the core of this book. We can make our cities more resilient by making them more beautiful, by making them more equitable, by restoring their ecological integrity. This is not a simple task, but neither is it too late in time nor too ambitious in reach. To the contrary, it is the ambition of an aim, set suitably high, that inspires us to the work.

References

Prologue

[1] Brazil may be the owner of 20% of the world's water supply but it is still very thirsty. The World Bank, August 5, 2016: www.worldbank.org/en/news/feature/2016/07/27/how-brazil-managing-water-resources-new-report-scd.

[2] Davies, W. Brazil drought: Sao Paulo sleepwalking into water crisis. *BBC News*, November 7, 2014: www.bbc.com/news/world-latin-america-29947965.

[3] Angelakis, A. Minoan aqueducts: A pioneering technology. 1st International Symposium on Water and Wastewater Technologies in Ancient Civilizations, Iraklio, Greece, 28–30 October 2006.

[4] Romero, S. A willing explorer of São Paulo's polluted rivers. *The New York Times*, December 14, 2012.

[5] Morcroft, G. Drought watch: One of Brazil's biggest cities has only 100 days of water supply left. *International Business Times*, July 29, 2014.

[6] Timerman, J. No one's quite sure how São Paulo will survive its drought. CityLab [*Bloomberg*], February 5, 2015.

[7] Watts, J. Brazil's worst drought in history prompts protests and blackouts. *The Guardian*, January 23, 2015.

[8] Philips, D. São Paulo faces critical water shortage as the World Cup prepares to kick off. *The Guardian*, May 21, 2014.

[9] Gerberg, J. A megacity without water: São Paulo's drought. *Time Magazine*, October 13, 2015.

[10] Zaitchik, A. Rainforest on fire. *The Intercept*, July 6, 2019.

[11] Gibbs, H., Munger, J., L'Roe, J., & Barreto, P. Did ranchers and slaughterhouses respond to zero-deforestation agreements in the Brazilian Amazon? Conservation Letters 9, 32–42 (2016).

[12] Mongabay. Mean net primary production by ecosystem: https://rainforests.mongabay.com/03net_primary_production.htm.

[13] Alexander, P. et al. Human appropriation of land for food: The role of diet. *Global Environmental Change* 41, 88–98 (2016).

[14] Cattle ranching in the Amazon region. Global Forest Atlas, Yale School of Forestry and Environmental Studies: https://globalforestatlas.yale.edu/amazon/land-use/cattle-ranching.

[15] Schmidinger, K. & Stehfast, E. Including CO_2 implications of land occupation in LCAs: Method and example for livestock products. *International Journal of Lifecycle Assessment* 17, 962–972 (2012).

[16] Butler, R. Ecology of the Amazon rainforest. Mongabay: https://rainforests .mongabay.com/amazon/rainforest_ecology.html.

[17] Nobre, P., Malagutti, M., Urbano, D., Almeida, R., & Giarolla, E. Amazon deforestation and climate change in a coupled model simulation. *Journal of Climate* 22, 5686–5697 (2009).

[18] Whately, M. & Lerer, R. Brazil drought: Water rationing alone won't save São Paulo. *The Guardian*. February 11, 2015.

[19] Vigna, A. Where has São Paulo's water gone? *Equal Times*, May 18, 2015.

[20] Watts, J. Brazil drought crisis leads to rationing and tensions. *The Guardian*, September 5, 2014.

[21] Watts, J. The Amazon effect: How deforestation is starving São Paulo of water. *The Guardian*, November 28, 2017.

[22] Brazil's southeastern states are experiencing the worst drought the country has seen in over 80 years. *Telesur*, February 19, 2015.

[23] Grabow, S. & Heskin, A. Foundations for a radical concept of planning. *Journal of the American Institute of Planners* 39, 106–114 (1973).

[24] Beard, V. Learning radical planning: The power of collective action. *Planning Theory* 2, 13–35 (2003).

[25] Miraftab, F. Insurgent planning: Situating radical planning in the global south. *Planning Theory* 8, 32–50 (2009).

[26] Jacobs, F. Black feminism and radical planning: New directions for disaster planning research. *Planning Theory* 18, 4–39 (2019).

[27] Williams, R. & Steil, J. The past we step into and how we repair it: A normative framework for reparative planning. Journal of the American Planning Association (2022): https://doi.org/10.1080/01944363.2022.2154247.

Chapter 1

[1] Dutta, A. Railway protection force estimates 80 deaths on Shramik trains. *Hindustan Times*, May 30, 2020.

[2] Kumar, A. India wilts as temperature hits 50 degrees Celsius. *Phys.org* (2021): https://phys.org/news/2020-05-india-wilts-heatwave-temperature-degrees.html.

[3] Masters, J. Death Valley, California, breaks the all-time world heat record for the second year in a row. *Yale Climate Connections*, July 12, 2021: https://yaleclimateconnections.org/2021/07/death-valley-california-breaks-the-all-time-world-heat-record-for-the-second-year-in-a-row/.

[4] Samenow, J. India's hellish heat wave, in hindsight. *The Washington Post*, June 10, 2015.

[5] Lim, C. L. Heat sepsis precedes heat toxicity in the pathophysiology of heat stroke: A new paradigm on an ancient disease. *Antioxidants* 7, 8–10 (2018).

[6]   Engelhardt, E. Apoplexy, cerebrovascular disease, and stroke: Historical evolu-
      tion of terms and definitions. *National Library of Medicine* 11, 449–453 (2017).

[7]   Jarcho, S. A Roman experience with heat stroke in 24 BC. *Bulletin of the
      New York Academy of Medicine* 43, 767–768 (1967).

[8]   Leithead, C. S. & Lind, A. R. *Heat Stress and Heat Disorders* (Cassell & Co.,
      1964).

[9]   Mufson, S. Facing unbearable heat, Qatar has begun to air-condition the outdoors.
      The Washington Post, October 16, 2019.

[10] Sherwood, S. C. & Huber, M. An adaptability limit to climate change due to heat
      stress. *Proceedings of the National Academy of Sciences of the United States of
      America* 107, 9552–9555 (2010).

[11] Raymond, C., Matthews, T., & Horton, R. M. The emergence of heat and humidity
      too severe for human tolerance. *Science Advances* 6 (2020): https://doi.org/10
      .1126/sciadv.aaw1838.

[12] Milman, O. 2020 was hottest year on record by narrow margin, NASA says. *The
      Guardian*, January 14, 2021.

[13] Florida Department of Health. 1998 wildland fire outbreak. Factsheet: https://
      flbrace.org/images/docs/wildland-fire-factsheet.pdf.

[14] Csongos, F. 1998 in review: A year of natural disasters. *RadioFreeEurope:
      RadioLiberty*, December 9, 1998.

[15] Hernandez, A. & McGinley, L. Harvard study estimates thousands died in Puerto
      Rico due to Hurricane Maria. *The Washington Post*, May 29, 2018.

[16] NOAA (National Oceanic and Atmospheric Association). 2017 Atlantic
      Hurricane Season. National Hurricane Center and Central Pacific Hurricane
      Center (2017): www.nhc.noaa.gov/data/tcr/index.php?season=2017&basin=atl.

[17] Ingraham, C. Houston is experiencing its third "500-year" flood in 3 years: How is
      that possible? *The Washington Post*, August 29, 2017.

[18] Schär, C. Climate extremes: The worst heat waves to come. *Nature Climate
      Change* 6, 128–129 (2016).

[19] Dunne, J. P., Stouffer, R. J., & John, J. G. Reductions in labour capacity from heat
      stress under climate warming. *Nature Climate Change* 3, 563–566 (2013).

[20] Cook-Anderson, G. Drought, urbanization were ingredients for Atlanta's perfect
      storm. *NASA*: www.nasa.gov/topics/earth/features/atlanta_tornado.html.

[21] Murray, B. On this date in 2008: The SEC basketball tournament tornado miracle
      shot. *AlabamaWX* (blog), March 14, 2017: www.alabamawx.com/?p=128876.

[22] Cimons, M. Humans have been messing with the climate for thousands of years.
      Popular Science, September 19, 2018.

[23] Stone, B. *The City and the Coming Climate: Climate Change in the Places We
      Live* (Cambridge University Press, 2012).

[24] Stone, B., Vargo, J., & Habeeb, D. Managing climate change in cities: Will
      climate action plans work? *Landscape and Urban Planning* 107, 263–271 (2012).

[25] Habeeb, D., Vargo, J., & Stone, B. Rising heat wave trends in large US cities.
      Natural Hazards 76, 1651–1665 (2015).

[26] Vaidyanathan, A., Malilay, J., Schramm, P., & Saha, S. Heat-related deaths:
      United States, 2004–2018. *CDC Morbidity and Mortality Weekly Report
      (MMWR)* 69, 729–734 (2020).

[27] Stone, B., Lanza, K., Mallen, E., Vargo, J., & Russell, A. Urban heat management in Louisville, Kentucky: A framework for climate adaptation planning. *Journal of Planning and Education and Research* 43, 346–358 (2019).

[28] Simpson, J. R. Urban forest impacts on regional cooling and heating energy use: Sacramento County case study. *Journal of Arboriculture* 24, 201–214 (1998).

[29] Turner-Skoff, J. & Cavender, N. The benefits of trees for livable and sustainable communities. *Plants, People, Planet* 1, 323–335 (2019).

[30] Ryu, Y. H., Bou-Zeid, E., Wang, Z. H., & Smith, J. A. Realistic representation of trees in an urban canopy model. *Boundary Layer Meteorology* 159, 193–220 (2016).

[31] Middel, A., Chhetri, N., & Quay, R. Urban forestry and cool roofs: Assessment of heat mitigation strategies in Phoenix residential neighborhoods. *Urban Forestry & Urban Greening* 14, 178–186 (2015).

[32] Zölch, T., Maderspacher, J., Wamsler, C., & Pauleit, S. Using green infrastructure for urban climate-proofing: An evaluation of heat mitigation measures at the micro-scale. *Urban Forestry & Urban Greening* 20, 305–316 (2016).

[33] Nowak, D. J. & Greenfield, E. J. Declining urban and community tree cover in the United States. *Urban Forestry & Urban Greening* 32, 32–55 (2018).

[34] Ziter, C. D., Pedersen, E. J., Kucharik, C. J. & Turner, M. G. Scale-dependent interactions between tree canopy cover and impervious surfaces reduce daytime urban heat during summer. *Proceedings of the National Academy of Sciences of the United States of America* 116, 7575–7580 (2019).

[35] O'Neil-Dunne, J. *A Report on the City of Cambridge's Existing and Possible Tree Canopy* (University of Vermont Spatial Analysis Laboratory, 2012): www.cambridgema.gov/-/media/Files/CDD/Climate/treecanopystudy/UVM_Tree_Study_20120807.pdf.

[36] O'Neil-Dunne, J. *A Report on the City of New York's Existing and Possible Tree Canopy* (University of Vermont Spatial Analysis Laboratory, 2010): www.fs.usda.gov/nrs/utc/reports/UTC_NYC_Report_2010.pdf.

[37] O'Neil-Dunne, J. *A Report on Washington, D.C.'s Urban Tree Canopy* (University of Vermont Spatial Analysis Laboratory, 2010): www.uvm.edu/~joneildu/Blog/Reports/UTC_Report_DC.pdf.

[38] O'Neil-Dunne, J. *A Report on the City of Philadelphia's Existing and Possible Tree Canopy* (University of Vermont Spatial Analysis Laboratory, 2011): www.phila.gov/media/20200210164446/Urban-Tree-Canopy-Report-03-18-11.pdf.

[39] O'Neil-Dunne, J. P. M. *A Report on the City of Baltimore's Existing and Possible Urban Tree Canopy* (University of Vermont Spatial Analysis Laboratory, 2009): http://gis.w3.uvm.edu/utc/Reports/TreeCanopy_Report_BACI_2007.pdf.

[40] Plan-It Geo. *Treasure Valley Urban Tree Canopy Assessment* (2013): www.cityofboise.org/media/4256/2013_treasure_valley_utc_project_report-fina_-071013.pdf.

[41] Ramsey, J. Tree canopy cover and potential in Portland, OR: A spatial analysis of the urban forest and capacity for growth. Master's thesis, Portland State University (2019): https://doi.org/10.15760/etd.6988.

[42] Davey Resource Group. *Urban Tree Canopy Assessment Sacramento, CA* (2018): www.cityofsacramento.org/-/media/Corporate/Files/Public-Works/Maintenance-Services/Urban-Forest-Master-Plan/Copy-of-Sacramento-UTC-Assessment-20180515.pdf?la=en.

[43] Song, X. P., Tan, P. Y., Edwards, P., & Richards, D. The economic benefits and costs of trees in urban forest stewardship: A systematic review. *Urban Forestry & Urban Greening* 29, 162–170 (2018).

[44] Kunsch, A. & Parks, R. *Tree Planting Cost-Benefit Analysis: A Case Study for Urban Forest Equity in Los Angeles* (TreePeople, 2021): www.treepeople.org /wp-content/uploads/2021/07/tree-planting-cost-benefit-analysis-a-case-study-for-urban-forest-equity-in-los-angeles.pdf.

[45] City of Cambridge Annual Budget 2021–2022 (2022): www.cambridgema.gov /-/media/Files/budgetdepartment/FinancePDFs/fy22submittedbudget/fy22sub mittedbudgetbook.pdf.

[46] Hotz, R. L. To offset climate change scientists tout city trees and ultra-white paint. Wall Street Journal, June 4, 2021.

[47] Labrie, S. Measuring how effectively the Toronto District School Board Management Plans have increased playground shade over a thirteen-year period. Master's thesis, University of Toronto (2017).

[48] US Centers for Disease Control and Prevention. *Shade Planning for America's Schools* (2008): www.cdc.gov/cancer/skin/pdf/shade_planning.pdf.

[49] Olsen, H., Kennedy, E., & Vanos, J. Shade provision in public playgrounds for thermal safety and sun protection: A case study across 100 play spaces in the United States. *Landscape and Urban Planning* 189, 200–211 (2019).

[50] Colter, K. R., Middel, A. C., & Martin, C. A. Effects of natural and artificial shade on human thermal comfort in residential neighborhood parks of Phoenix, Arizona, USA. *Urban Forestry & Urban Greening* 44 (2019): https://doi.org/10 .1016/j.ufug.2019.126429.

[51] Johansson, E. & Emmanuel, R. The influence of urban design on outdoor thermal comfort in the hot, humid city of Colombo, Sri Lanka. *International Journal of Biometeorology* 51, 119–133 (2006).

[52] Watanabe, S. & Ishii, J. Effect of outdoor thermal environment on pedestrians' behavior selecting a shaded area in a humid subtropical region. *Building and Environment* 95, 32–41 (2016).

[53] Martinelli, L., Lin, T. P., & Matzarakis, A. Assessment of the influence of daily shadings pattern on human thermal comfort and attendance in Rome during summer period. *Building and Environment* 92, 30–38 (2015).

[54] Shashua-Bar, L., Pearlmutter, D., & Erell, E. The influence of trees and grass on outdoor thermal comfort in a hot-arid environment. *International Journal of Climatology* 31, 1498–1506 (2011).

[55] Alvey, A. A., Wiseman, P. E., & Kane, B. Efficacy of conventional tree stabilization systems and their effect on short-term tree development. *Arboriculture & Urban Forestry* 35, 157–164 (2009).

[56] Krayenhoff, E. S., Broadbent, A., Zhao, L., et al. Cooling hot cities: A systematic and critical review of the numerical modelling literature. *Environmental Research Letters*, 16 (2021): https://doi.org/10.1088/1748-9326/abdcf1.

[57] Santamouris, M., Synnefa, A., & Karlessi, T. Using advanced cool materials in the urban built environment to mitigate heat islands and improve thermal comfort conditions. *Solar Energy* 85, 3085–3102 (2011).

[58] Li, X., Peoples, J., Yao, P., & Ruan, X. Ultrawhite $BaSO_4$ Paints and Films for Remarkable Daytime Subambient Radiative Cooling. *ASC Applied Materials and Interfaces* 13, 21733–21739 (2021).

[59] Kumar, R., Mishra, V., Buzan, J., et al. Dominant control of agriculture and irrigation on urban heat island in India. *Scientific Reports* 7 (2017): https://doi.org/10.1038/s41598-017-14213-2.

[60] Lu, Y. & Kueppers, L. Increased heat waves with loss of irrigation in the United States. *Environmental Research Letters* 10 (2015): https://doi.org/10.1088/1748-9326/10/6/064010.

[61] Gao, K. & Santamouris, M. The use of water irrigation to mitigate ambient overheating in the built environment: Recent progress. *Building and Environment* 164 (2019): https://doi.org/10.1016/j.buildenv.2019.106346.

[62] Broadbent, A. M., Coutts, A. M., Tapper, N. J., & Demuzere, M. The cooling effect of irrigation on urban microclimate during heat wave conditions. *Urban Climate* 23, 309–329 (2018).

[63] Hendel, M., Gutierrez, P., Colombert, M., Diab, Y., & Royon, L. Measuring the effects of urban heat island mitigation techniques in the field: Application to the case of pavement-watering in Paris. *Urban Climate* 16, 43–58 (2016).

[64] Maillard, P., David, F., Dechesne, M., Bailly, J., & Lesueur, E. Characterization of the Urban Heat Island and evaluation of a road humidification solution in the district of La Part-Dieu, Lyon (France). *Techniques Sciences Méthodes* 6, 23–35 (2014).

[65] Santamouris, M., Ding, L., Fiorito, F., et al. Passive and active cooling for the outdoor built environment: Analysis and assessment of the cooling potential of mitigation technologies using performance data from 220 large scale projects. *Solar Energy* 154, 14–33 (2017).

Chapter 2

[1] Frank, T. After a $14-billion upgrade, New Orleans' levees are sinking. *Scientific American*, April 11, 2019.

[2] Ganuchcau, J. New Orleans flood August 2017. *Climate Signals*, December 4, 2018: www.climatesignals.org/events/new-orleans-flood-august–2017.

[3] Evans, B. 46 tons of Mardi Gras beads found in clogged catch basins. *The Times-Picayune/The New Orleans Advocate*, January 25, 2018.

[4] Craig, T. It wasn't even a hurricane but heavy rains flooded New Orleans as pumps faltered. *The Washington Post*, August 9, 2017.

[5] List of capital cities by elevation. Wikipedia (2022): https://en.wikipedia.org/wiki/List_of_capital_cities_by_elevation.

[6] Dahl, K. A., Fitzpatrick, M. F., & Spanger-Siegfried, E. Sea level rise drives increased tidal flooding frequency at tide gauges along the U.S. East and Gulf Coasts: Projections for 2030 and 2045. *PLOS One* 12 (2017): https://doi.org/10.1371/journal.pone.0170949.

[7] National Park Service. The Potowmack Canal: www.nps.gov/grfa/learn/history culture/canal.htm.

[8] Prillaman, M. Are we in the Anthropocene? Geologists could define new epoch for Earth. *Nature* 623, 14–15 (2023).

[9] Intergovernmental Panel on Climate Change (IPCC). The Physical Science Basis. In V. Masson-Delmotte, P. Zhai, A. Pirani et al. (eds.), Contribution of Working Group I to the Sixth Assessment Report of the Intergovernmental Panel on Climate Change (Cambridge University Press, 2021).

[10] Slater, T., Lawrence, I., Otosaka, I., et al. Review article: Earth's ice imbalance. *Cryosphere* 15, 233–246 (2021).

[11] Chen, X., Zhang, X., Church, J., et al. The increasing rate of global mean sea-level rise during 1993–2014. *Nature Climate Change* 7, 492–495 (2017).

[12] Shultz, K., Nguyen, T., & Pillar, H. An improved and observationally-constrained melt rate parameterization for vertical ice fronts of marine terminating glaciers. *Geophysical Research Letters* 49, e2022GL100654 (2022).

[13] NASA. IPCC 6th Assessment Report Sea Level Projection Tool: https://sealevel .nasa.gov/ipcc-ar6-sea-level-projection-tool?overlay_open=true.

[14] Bott, L. M., Schon, T., Illigner, J., et al. Land subsidence in Jakarta and Semarang Bay: The relationship between physical processes, risk perception, and household adaptation. *Ocean & Coastal Management* 211 (2021): https:// doi.org/10.1016/j.ocecoaman.2021.105775.

[15] Dejong, B., Bierman, P., Newell, W., et al. Pleistocene relative sea levels in the Chesapeake Bay region and their implications for the next century. *Geological Society of America Today* 25, 4–10 (2015).

[16] Tabari, M. Climate change impact on flood and extreme precipitation increases with water availability. *Scientific Reports* 10, 13768 (2020).

[17] Water Science School. How much water is there on earth? USGS, November 13 (2019): www.usgs.gov/special-topic/water-science-school/science/how-much-water-there-earth?qt-science_center_objects=0#qt-science_center_objects.

[18] Papalexiou, S. M. & Montanari, A. Global and regional increase of precipitation extremes under global warming. *Water Resources Research* 55, 4901–4914 (2019).

[19] Janssen, E., Wuebbles, D. J., Kunkel, K. E., Olsen, S. C., & Goodman, A. Observational- and model-based trends and projections of extreme precipitation over the contiguous United States. *Earth's Future* 2, 99–113 (2014).

[20] Fischer, E. M. & Knutti, R. Observed heavy precipitation increase confirms theory and early models. *Nature Climate Change* 6, 986–991 (2016).

[21] Myhre, G. et al. Frequency of extreme precipitation increases extensively with event rareness under global warming. *Scientific Reports* 9, 2–11 (2019).

[22] Express News Service. Gulp it down: Chennai rainfall among highest in last 200 years! *The New Indian Express*, November 28, 2021.

[23] Luu, L. N., Scussolini, P., Kew, S., et al. Attribution of typhoon-induced torrential precipitation in Central Vietnam, October 2020. *Climatic Change* 169, 1–22 (2021).

[24] Hollingsworth, J. More than 100 dead as Vietnam reels from "worst floods in decades." *CNN*, October 21, 2020.

[25] English, E. C., Li, M., Zarins, R., & Feltham, T. H. H. The economic argument for amphibious retrofit construction. Paper presented at the 8th International Conference on Building Resilience – ICBR, Lisbon, November 14–16, 2018.

[26] Busscher, T., van den Brink, M., & Verweij, S. Strategies for integrating water management and spatial planning: Organising for spatial quality in the Dutch "Room for the River" program. *Journal of Flood Risk Management* 12, 1–12 (2019).

[27] ClimateWire. How the Dutch make "Room for the River" by redesigning cities. *Scientific American*, January 20, 2012.

[28] Oosthoek, K. J. Dutch river defenses in historical perspective. Environmental History Resources (2006): www.eh-resources.org/dutch-river-defences-in-historical-perspective/#_ednref6.

[29] Rubin, S. In Amsterdam, a community of floating homes shows the world how to live alongside nature. *The Washington Post*, December 17, 2021.

[30] Greenprint: Lessons learned. Schoonschip: https://greenprint.schoonschipamsterdam.org/impactgebieden/sociaal#lessons.

[31] Hartman, K. J., Ap, L., Stieve, D. R., & Associates, E. *Floodproofing New York: The City's Response to Superstorm Sandy* (International Institute of Building Enclosure Consultants, 2018): https://iibec.org/wp-content/uploads/2018-cts-hartman-stieve.pdf.

[32] Parkinson, R. W. Speculation on the role of sea-level rise in the tragic collapse of the Surfside condominium (Miami Beach, Florida U.S.A.) was a bellwether moment for coastal zone management practitioners. *Ocean & Coastal Management* 215, 105968 (2021).

[33] Clark, J. Premium Elevation, LLC. *Connection* 19, 18–20 (2021).

[34] Hanks, D. & Ortiz, R. Will Dade enact a 60-day notice for rent increases over 5%? *Miami Herald*, February 23, 2022.

[35] Tan, R., Kornfield, M., & Brice-Saddler, M. In Miami's gentrifying neighborhoods, Surfside condo collapse deepens fears of displacement. *The Washington Post*, July 17, 2021.

[36] Hillier, A. E. Redlining and the Home Owners' Loan Corporation. *Journal of Urban History* 29, 394–420 (2003).

[37] Wilson, B. Urban heat management and the legacy of redlining. *Journal of the American Planning Association* 86, 443–457 (2020).

[38] Lukes, D. & Cleveland, C. The lingering legacy of redlining on school funding, Diversity, and Performance. EdWorkingPaper No. 21-363, 21–363 (2021).

[39] Capps, K. & Cannon, C. Redlined, now flooding. *Bloomberg*, March 15, 2021.

[40] Hoffman, J. S., Shandas, V., & Pendleton, N. The effects of historical housing policies on resident exposure to intra-urban heat: A study of 108 US urban areas. *Climate* 8, 1–12 (2020).

[41] Winship, S., Pulliam, C., Shiro, A., Reeves, R., & Deambrosi, S. *Long Shadows: The Black–White Gap in Multigeneration Poverty* (The Brookings Institution, 2021): www.brookings.edu/research/long-shadows-the-black-white-gap-in-multigenerational-poverty/.

[42] Lavizzo-Mourey, R. & Williams, D. Being black is bad for your health. *US News & World Report*, April 14, 2016.

[43] Garrison, J. D. Seeing the park for the trees: New York's "Million Trees" campaign vs. the deep roots of environmental inequality. *Environment and Planning B: Urban Analytics and City Science* 46, 914–930 (2019).

[44] The White House. Fact Sheet: President Biden Takes Executive Actions to Tackle the Climate Crisis at Home and Abroad, Create Jobs, and Restore Scientific Integrity Across Federal Government, January 27. www.whitehouse.gov/briefing-room/state ments-releases/2021/01/27/fact-sheet-president-biden-takes-executive-actions-to-tackle-the-climate-crisis-at-home-and-abroad-create-jobs-and-restore-scientific-integrity-across-federal-government/#:~:text=Today%2C%20President%20Biden% 20will%20take,the%20federal%20government%2C%20and%20re%2D.

[45] Murray-Cooper, A. Justice in Urban Climate Plans: How and Where Cities Are Integrating Equity and Climate. Boston University Initiative on Cities (2021). www.bu.edu/ioc/2021/11/01/justice-in-urban-climate-plans-how-and-where-cities-are-integrating-equity-and-climate/.

[46] City of Providence's Climate Justice Plan (2019): www.providenceri.gov/wp-content/uploads/2019/10/Climate-Justice-Plan-Report-FINAL-English-1.pdf.

Chapter 3

[1] Radtke Russell, P. Threats of "Day Zero" water scarcity multiply. *Engineering News-Record*, July 20, 2020.

[2] Parvatam, S. & Priyadarshini, S. On Day Zero, India prepares for a water emergency. *Nature India*, June 30, 2019.

[3] Oppili, P. 80% of Chennai was wetland in 1980s, now 15%. *The Times of India*, September 5, 2016.

[4] C40. The 2015 C40 Cities Award Winner's Circle. C40 Cities (blog), June 2, 2016: www.c40.org/blog_posts/the-2015-c40-cities-award-winners-circle.

[5] Prabahakaran, S. & Sule, T. Cape Town: Towards a sustainable water future. *Urban Water Atlas*, May 9, 2020: www.urbanwateratlas.com/2020/05/09/cape-town-towards-a-sustainable-water-future/.

[6] Alexander, C. Cape Town's "Day Zero" water crisis: One year later. *Bloomberg*, April 12, 2019.

[7] Zilber, A. NEWBrew beer in Singapore is made from recycled sewage water. *New York Post*, June 30, 2022.

[8] VOANews, Singapore turns sewage into clean, drinkable water, meeting 40% of demand. *VOANews*, August 10, 2021.

[9] BBC World News. The 11 cities most likely to run out of drinking water – like Cape Town. *BBC News*, February 11, 2018.

[10] Chiang, F., Mazdiyasni, O., & AghaKouchak, A. Evidence of anthropogenic impacts on global drought frequency, duration, and intensity. *Nature Communications* 12 (2021): https://doi.org/10.1038/s41467-021-22314-w.

[11] Xu, L., Chen, N., & Zhang, X. Global drought trends under 1.5 and 2°C warming. *International Journal of Climatology* 39, 2375–2385 (2019).

[12] Carrington, D. Climate limit of 1.5°C close to being broken, scientists warn. *The Guardian*, May 9, 2022.

[13] Sognnaes, I., Gambhir, A., Van de ven, D., et al. A multi-model analysis of long-term emissions and warming implications of current mitigation efforts. *Nature Climate Change* 11, 1055–1062 (2021).

[14] United Nations Environment Program. *Emissions Gap Report 2021* (2021): www
      .unep.org/resources/emissions-gap-report-2021.
[15] Nowak, D. J. & Greenfield, E. J. Tree and impervious cover change in U.S. cities.
      Urban Forestry & Urban Greening 11, 21–30 (2012).
[16] US Environmental Protection Agency (EPA). Water efficiency for water sup-
      pliers: Water efficiency strategies (2023): www.epa.gov/sustainable-water-
      infrastructure/water-efficiency-water-suppliers.
[17] Los Angeles County Waterworks Districts. Drought frequently asked questions:
      https://dpw.lacounty.gov/wwd/web/Conservation/droughtinfoFAQ.aspx.
[18] Brulliard, K. The Colorado River is in crisis, and it's getting worse every day. *The
      Washington Post*, May 14, 2022.
[19] Slivka, K. A Mayan water system with lessons for today. *The New York Times*,
      July 16, 2012.
[20] Teston, A., Geraldi, M. S., Colasio, B. M. & Ghisi, E. Rainwater harvesting in
      buildings in Brazil: A literature review. *Water* 10, 471 (2018).
[21] Campisano, A., Butler, D., Ward, S., et al. Urban rainwater harvesting systems:
      Research, implementation and future perspectives. *Water Research* 115, 195–209
      (2017).
[22] Pushkarna, V. Rainwater harvesting: Delhi's focus moves from rooftops to public
      parks and other open spaces. *Citizen Matters*, July 12, 2022.
[23] Han, M. Progress of multipurpose and proactive rainwater management in Korea.
      Environmental Engineering Research 18, 65–69 (2013).
[24] US Environmental Protection Agency (EPA). *Stormwater Best Management
      Practices: Permeable Pavements.* Report no. EPA-832-F-21-031W (EPA, 2021).
[25] Amol, P., Priyanka, K., & Onkar, S. *Green Roof Market Size, Share & Trends
      Analysis Report by Type, by Application, by Region, and Segment Forecasts,
      2020–2027.* Report no. GVR-3-68038-183-2 (Grand View Research, 2020):
      www.grandviewresearch.com/industry-analysis/green-roof-market.
[26] United States Congress. Infrastructure Investment and Jobs Act. 117–58 (2021).
[27] Seliktar, O. Turning water into fire: The Jordan River as the hidden factor in the
      Six Day War. *Middle East Review of International Affairs*, 9: 57–71 (2005).
[28] Novo, C. Israel leads the way in wastewater reuse. *Smart Water Magazine*,
      July 29, 2020.
[29] Rabinovitch, A., Barrington, L., & Al-Khalidi, S. Israel, Jordan to partner in
      water-for-energy deal. *Reuters*, November 8, 2021.
[30] Ritchie, H. & Roser, M. Israel: Energy country profile. Our World in Data: https://
      ourworldindata.org/energycountry/israel#how-much-of-the-country-s-energy-
      comes-from-fossil-fuels.
[31] Ding, J. Los Angeles could soon put recycled water directly in your tap. It's not
      "toilet to tap." *Los Angeles Times*, July 22, 2022.
[32] Alawadhi, A. & Tartakovsky, D. M. Bayesian update and method of distributions:
      Application to leak detection in transmission mains. *Water Resources Research*
      56, 1–10 (2020).
[33] Blount, S. Home water use in the United States. National Environmental
      Education Foundation: www.neefusa.org/weather-and-climate/weather/
      home-water-use-united-states.

[34] European Environment Agency. Water use in Europe: Quantity and quality face big challenges (2018): www.eea.europa.eu/signals/signals-2018-content-list/art icles/water-use-in-europe-2014.

[35] Luby, I. H., Polasky, S., & Swackhamer, D. L. US urban water prices: Cheaper when drier. *Water Resources Research* 54, 6126–6132 (2018).

[36] Parks, R., Mclaren, M., Toumi, P. R., & Rivett, P. U. Experiences and lessons in managing water from Cape Town. London Imperial College, Grantham Institute Briefing Paper No. 29 (2019): www.imperial.ac.uk/media/imperial-college/gran tham-institute/public/publications/briefing-papers/Experiences-and-lessons-in-managing-water.pdf.

[37] Shapiro, A. Cape Town averts "Day Zero" by limiting water use. *National Public Radio*, June 28, 2018.

[38] Urban Sustainability Exchange, City of Cape Town's Water Map: https://use.metropolis.org/case-studies/city-of-cape-towns-water-map.

Chapter 4

[1] Yoon, J. Ian becomes a hurricane again as it takes aim at South Carolina. *The New York Times*, September 29, 2022.

[2] Stebbins, S. Cape Coral is one of the fastest growing cities in America. 24/7 Wall St (2021): https://247wallst.com/city/cape-coral-is-one-of-the-fastest-growing-cities-in-america/.

[3] Victor, D. Hurricane Andrew: How *The Times* reported the destruction of 1992. *The New York Times*, September 6, 2017.

[4] National Hurricane Center. Tropical cyclone naming history and retired names: www.nhc.noaa.gov/aboutnames_history.shtml.

[5] Tavernise, S. Did Hurricane Ian bust Florida's housing boom? *The New York Times*, October 18, 2022.

[6] Allen, G. Hurricane Katrina: 10 years of recovery and reflection ghosts of Katrina still haunt New Orleans' shattered Lower Ninth Ward. *National Public Radio*, August 3, 2015.

[7] Flavelle, C. Hurricane Ian's toll is severe. Lack of insurance will make it worse. *The New York Times*, September 9, 2022.

[8] FEMA. Moving out of harm's way proves advantageous and gives rise to widely used park (2021): www.fema.gov/case-study/moving-out-harms-way-proves-advantageous-and-gives-rise-widely-used-park.

[9] PEW. Repeatedly flooded properties cost billions. Infographic (2016): www.pewtrusts.org/-/media/assets/2016/10/repeatedly_flooded_properties_cost_bil lions.pdf.

[10] Painter, W. L. *The Disaster Relief Fund: Overview and Issues* (Congressional Research Services, 2022): https://sgp.fas.org/crs/homesec/R45484.pdf.

[11] US Government Accountability Office. Flood mitigation: Actions needed to improve use of FEMA property acquisitions. September (2022): www.gao.gov/products/gao-22-106037.

[12] Hauer, M. E., Evans, J. M., & Mishra, D. R. Millions projected to be at risk from sea-level rise in the continental United States. *Nature Climate Change* 6, 691–695 (2016).

[13] Pinter, N. True stories of managed retreat from rising waters. *Issues in Science and Technology* 37, 64–73 (2021).

[14] US National Park Service. Native American Heritage of the Niobrara (2020): https://www.nps.gov/niob/native-american-heritage-of-the-niobrara.htm.

[15] Davis, M. Cahokia: North America's massive, ancient city. Big Think (2023): https://bigthink.com/the-past/cahokia/.

[16] Seppa, N. Metropolitan life on the Mississippi. The Washington Post, March 12, 1997.

[17] Newitz, A. *Four Lost Cities: A Secret History of the Urban Age* (W. W. Norton & Company, 2021).

[18] Cottar, S., Doberstein, B., Henstra, D., & Wandel, J. Evaluating property buyouts and disaster recovery assistance (rebuild) options in Canada: A comparative analysis of Constance Bay, Ontario and Pointe Gatineau, Quebec. *Natural Hazards* 109, 201–220 (2021).

[19] Donovan, L. Buyouts on the ballot: 10 years after Sandy, New York considers new funding for voluntary relocation. *City Limits*, October 24, 2022.

[20] New York Proposal 1, Environmental Bond Measure. Ballotpedia (2022): https://ballotpedia.org/New_York_Proposal_1,_Environmental_Bond_Measure_ (2022).

[21] Alexander, K. Reclaiming paradise. San Francisco Chronicle, May 3, 2019.

[22] Siegler, K. In fire scorched California, town aims to buy the highest at-risk properties. *National Public Radio*, August 23, 2021.

[23] Agyeman, J., Devine-Wright, P., & Prange, J. Close to the edge, down by the river? Joining up managed retreat and place attachment in a climate changed world. *Environment and Planning A: Economy and Space* 41, 509–513 (2009).

[24] Galway, L. P., Beery, T., Jones-Casey, K., & Tasala, K. Mapping the solastalgia literature: A scoping review study. *International Journal of Environmental Research and Public Health* 16 (2019): https://doi.org/10.3390/ijerph16152662.

[25] US Department of the Interior. Biden–Harris administration makes $135 million commitment to support relocation of tribal communities affected by climate change. (2022): www.doi.gov/pressreleases/biden-harris-administration-makes-135-million-commitment-support-relocation-tribal.

[26] Georgetown Climate Center. *Managing the Retreat from Rising Seas* (Georgetown Climate Center, 2020): www.georgetownclimate.org/files/MRT/GCC_20_FULL-3web.pdf.

[27] Kulp, S. A. & Strauss, B. H. New elevation data triple estimates of global vulnerability to sea-level rise and coastal flooding. *Nature Communications* 10 (2019): https://doi.org/10.1038/s41467-019-12808-z.

[28] Flavelle, C. Trump administration presses cities to evict homeowners from flood zones. *The* New York Times, March 11, 2020.

[29] Semon, N. J. Punta Gorda tallying damage from Hurricane. *YourSun.com*, October 10, 2022: www.yoursun.com/storm/latest_weather/punta-gorda-tallying-damage-from-hurricane-ian/article_fdbbd38c-4417-11ed-8721-33955a307d6e.html.

[30] Zambrano, L. Cape Coral residents weather Hurricane Ian: 'It's been a catastrophic event for the city'. Fort Myers News-Press, September 29, 2022.

[31] US Environmental Protection Agency (EPA). Fast facts on transportation greenhouse gas emissions (2022): www.epa.gov/greenvehicles/fast-facts-transportation-greenhouse-gas-emissions.

[32] Plumer, B. Cars take up way too much space in cities. New technology could change that. *Vox* (2016): www.vox.com/a/new-economy-future/cars-cities-technologies.

[33] Pruetz, R. Ecocity Snapshots: Nijmegen Netherlands – Room for the River. *Ecocities Emerging* (2017): https://ecocitiesemerging.org/nijmegen-netherlands-room-for-the-river/.

Postscript

[1] Faulkner, T. Before and after: Innovative design fights flooding in Atlanta. *Trust for Public Land* (2023): www.tpl.org/stories/cook-park-atlanta-green-infrastructure-flooding.

[2] Atlanta Regional Commission (ARC). *Neighborhood Statistical Area L01 Fact Sheet* (2023): http://documents.atlantaregional.com/NN/Profiles/AtlantaProfiles/L01.pdf.

[3] Stevens, A. Study: 4 Atlanta neighborhoods among nation's most dangerous. Atlanta Journal Constitution, October 5, 2010.

[4] Wang, Y., Shen, J. K., Xiang, W., & Wang, J. Q. Identifying characteristics of resilient urban communities through a case study method. *Journal of Urban Management* 7, 141–151 (2018).

[5] Saporta, M. Westside Future Fund offers lessons on building communities from the ground up. Saporta Report, January 23, 2023.

[6] Green, J. Five years after Mercedes-Benz Stadium broke ground, is Atlanta's Westside revival working? *Curbed Atlanta*, January 31, 2019.

[7] Park Pride. *Proctor Creek North Avenue Watershed Basin: A Green Infrastructure Vision* (Park Pride, 2010). https://parkpride.org/wp-content/uploads/2016/09/2010_pna_overview-1.pdf.

[8] AgLanta. *Fresh Food Access Report City of Atlanta 2020* (Department of City Planning and AgLanta, 2020): www.aglanta.org/2020-fresh-food-access-report.

Index